The Money-Saving Garden Year

A month-by-month guide to a great garden that costs less

ANYA LAUTENBACH

CONTENTS

INTRODUCTION

Welcome to my guide to saving money in your garden throughout the year. In this book, I show you how every day provides an opportunity to create something new that costs very little and gives you a great sense of achievement. Taking you through each month, I offer you my 35 years' experience of planting and propagating, so that you can share the joy I feel when I see roots developing on cuttings and my beds filled with flowers. There's something for everyone, too, no matter what size your outdoor space, be it a few pots on a windowsill or a large garden, and whether you have been gardening for years or are just starting your adventure. Let me take you on a journey.

Through the advice in these pages, I will help you to create the garden of your dreams without spending much money. It's not about doing things on a budget, but simply working in harmony with Nature to produce spectacular, magical gardens and breath-taking outdoor spaces, all at a fraction of the usual cost. With my tips you'll be amazed by what you can achieve.

For me, plants are like old friends, forgiving and always there for us when we need them every day of the year. I love how everything is always changing and evolving in my garden, right before my eyes. Nothing ever stays the same, which means I never get bored. I lose myself in the never-ending beauty that comes in waves with each season: the freshness of spring, vibrancy of summer, rich colours of autumn, and the peace of winter. And when I go through challenging and difficult times, the garden also reminds me of the ebb and flow of life, offering hope for the future – like our plants, which are rejuvenated and come back stronger after a period of dormancy, we will bounce back in time.

Many of you will know that I have attention deficit hyperactivity disorder (ADHD) and in each month's Mindful Moments I also share with you the ways gardening helps me to focus and stay connected to the world around me. I also love wildlife and want to draw your attention to some of the creatures that are essential to a garden's healthy ecosystem but are often overlooked or even killed with pesticides. I show how some can actually save you money, too, and by nurturing wildlife you will also be helping to restore the planet while making a beautiful, eco-friendly garden you can be proud of.

As well as sharing my seasonal tips of what to grow and propagate each month and how to maintain your beautiful space, I have also thrown a spotlight on some of my favourite plants, such as alliums, roses, lavender, and dahlias, in the In Focus pages, and included simple projects for you to try.

In addition, I've put together my essential, easy-to-use Plant Propagation Guide at the back of the book, detailing the timings and methods of growing a wide range of plants for free, from sowing flower seeds to taking cuttings from trees and shrubs.

While we might not know each other very well, I have a feeling that we have a lot in common. The fact that you are holding this book in your hand today and reading my words tells me that you see the same magic in this world that I see, or perhaps you are searching for a way to learn about it. Well, you are definitely in the right place, so turn the page and enjoy.

CHAPTER 1

JANUARY

It's midwinter and the garden may be looking a bit forlorn, but I always find some flowers to enjoy, the blooms sparkling like jewels on cold January days. Many shrubs that flower now offer a feast for our senses, too, with sweet box (*Sarcococca*) and witch hazel (*Hamamelis*) emitting a wonderful fragrance that helps me to reconnect with Nature.

Evergreens such as *Euonymus fortunei* and *Pittosporum* add to the colourful palette, while the soft catkins of goat willow (*Salix caprea*) just demand to be touched. My mum used to cut willow stems and bring them into the house, and I remember as a child being fascinated by the furry catkins – a memory I like to recreate in my own home in January.

I also enjoy my houseplants at this time of year. Amaryllis comes into flower at the end of the month, and the green leaves of the sword fern (*Nephrolepis exaltata*) and *Aloe vera* make the house feel like an indoor garden.

IN SEASON

I make a special effort to keep the garden looking good in January. My spirits can dip as winter's dark days stretch before me and the weather keeps me indoors, but I take heart in the natural beauty to be found at this time of year.

WHAT'S IN BLOOM?

You may think that January has little to offer in terms of flower colour and interest, but there are a surprising number of plants that put on a performance now to bring cheer.

Euphorbia characias
An evergreen that sparkles in the winter garden when dusted in frost. The variegated Silver Swan is particularly eye-catching.

Prunus × *subhirtella* **'Autumnalis Rosea'**
This small cherry tree flowers throughout the winter months, with a sprinkling of pale pink flowers appearing intermittently from late autumn through winter to early spring.

Sweet box (*Sarcococca*)
A beautiful evergreen winter-flowering shrub, sweet box produces tiny but highly scented white flowers, and grows well in dry shade.

Mahonia x *media*
Another of my favourite winter-flowering shrubs, this beauty has large prickly evergreen leaves and arched spikes of sweetly scented yellow winter flowers, loved by pollinators.

Witch hazel (*Hamamelis*)
One of the stars of the winter garden, the spidery blooms of witch hazel scent the air with their amazing perfume. This large shrub needs acid soil to thrive so check yours with a soil pH kit, available from the garden centre, before buying one or planting a cutting.

MINDFUL MOMENT

I have my own ways of staying sane in January, which I've done all my life to keep me in balance. First, I spend as much time as possible outside – gardening, walking, or running. I also like to propagate plants now. I've been doing this since I was a little girl and it is my own very special medicine. When I see new roots on cuttings, I feel so excited by the fact that I'm creating new life and I advise everyone to have a go and see how it makes them feel. It's a great way to stay in touch with nature.

SPOTTING WILDLIFE

Common Woodlouse
Many people think this little creature is a pest, but, in fact, the woodlouse plays a vital role in recycling nutrients for plants to absorb, while doing them no harm at all. Woodlice are not actually insects but crustaceans and related to crabs and lobsters. They seek out dark, damp areas of the garden and eat dead wood, fallen leaves, fungi, and fruit, which they munch on and then deliver a neat packet of plant nutrients back to the soil through their faeces. It is, therefore, important not to kill them or use pesticides that may destroy your very own army of composters.

Clockwise from top left: witch hazel (*Hamamelis*) produces spidery, scented blooms in winter; many hellebores start to flower in January; I take hardwood cuttings of my colourful dogwoods (*Cornus*) now; *Euphorbia characias* Silver swan looks beautiful dusted in snow in winter – grow it in free-draining soil and a spot in full sun.

Clockwise from top left: I propagate my roses from hardwood cuttings taken in the winter months; when the weather keeps me indoors, I love nothing better than sorting through the seeds I will be sowing when the weather warms up; use this month to clean and varnish garden tools; plant dahlias now if you have space indoors.

WHAT TO GROW IN JANUARY

While this is one of the quietest months in the garden, I can always find jobs to keep me busy in January, ordering seeds, taking a few cuttings and planting bare-root trees and shrubs, which are usually cheaper than those grown in pots.

SOW INDOORS

Flowers

There are a few plants that take a while to germinate and establish from seed, which I like to sow early in the year. Keep the following in a warm place with plenty of sunlight for the best results.

- Sweet peas (*Lathyrus odoratus*)
- Hardy annuals such as *Cerinthe* and pot marigolds (*Calendula*)
- Annual climbers, including the cup and saucer plant (*Cobaea scandens*) and black-eyed Susans (*Thunbergia alata*)

Herbs, vegetables, and fruit

- Microgreens, which can be sown on a windowsill all year round.

PLANT OUTSIDE

The best time to buy new trees, shrubs, hedging plants, and some climbers, is during the winter months, when they are dormant.

Nurseries grow these plants in open fields, then dig them up after the leaves have fallen in autumn. They then sell them as bare-root plants, which are basically the stems and rootballs, with no soil around them. You will find them at supermarkets, garden centres, and online nurseries at a fraction of the cost of those raised in containers.

Bare-root trees and shrubs

Plant these as soon as they arrive, if the soil is not frozen or waterlogged, making sure that they are at the same level they were growing at in the soil, which you can see from the darker mark on the lower trunk or stems.

Bare-root roses

These are also available now to be planted at the same depth as they were in the soil. Just make sure that the graft union (knobbly area on the lower stems) is at soil level and not buried beneath the surface.

Hedging plants

You can propagate hedging from existing plants (see p.100) and plant them now, or buy small bare-root plants from a nursery.

PROPAGATE NOW

Take hardwood cuttings

Most shrubs and roses grow successfully from hardwood cuttings taken in the winter months from December to February (see p.156).

Pot up dahlia tubers

Plant the finger-like tubers in large pots, about 13cm (5in) deep, in peat-free compost and keep them indoors. You will then be able to take cuttings from the new shoots in spring.

EASY MONEY-SAVING IDEA

I like to save money and repurpose items that may go into landfill by making my own biodegradable seed pots from cardboard tubes, old magazines, and plant leaves. I have a little pot-making press, which I use to create paper pots, and last year I also tried using large leaves, which worked well, too. You can also repurpose food packaging such as yoghurt pots (make drainage holes in the bottom) rather than throwing them away.

MAINTENANCE MUSTS

Clean the greenhouse

Accomplishing a task outside always brings much-needed positivity in the darker months of the year. I use these dull days to have a general tidy-up of the greenhouse or other areas of the garden, such as the patio or pots outside our front door.

Prune fruit trees such as apples and pears

Take out any dead, diseased and crossing stems, and a few branches from the centre of the tree to allow more light to reach the fruits later in the year. Do not prune cherries or other members of the *Prunus* family – wait until summer when the fungal diseases that affect them are less prevalent. Try using your prunings to make plant supports.

Remove old leaves from hellebores

Remove hellebores' old leaves between autumn and winter to give the flowering stems a better chance to flower. Their old and damaged leaves don't look good at this time of the year anyway and pruning them out allows more light to reach the flowering stems, while new leaves will replace the old ones in the spring. Cutting off last year's foliage also prevents hellebore leaf spot disease from persisting through the winter on the leaves and infecting the flowers.

Clean and varnish tools

Use the winter months to clean and sharpen your gardening tools, ready for spring. I also take this opportunity to varnish the wooden handles to keep them in good condition.

Prune roses

It's a good time to focus on your roses now. If you have bush roses, cut down one or two old stems close to the ground, and shorten the rest by between a third and a half. Climbing roses can be pruned by taking one or two of the oldest stems back to the base of the plant, then cut back any side-shoots that produced flowers last year by two-thirds of their length. Remember to also tie any new growth to its support. Renovate ramblers in winter, too, by removing the older stems to leave about six or seven young, healthy stems and shorten the side-shoots by one-third to one-half. Leave other rambling roses, which are best pruned in summer, after flowering.

MONEY-SAVING PROPAGATION TIP

When pruning your shrubs, don't simply throw out or compost your prunings. You can use some to take hardwood cuttings (see p.156) and get more plants for free for your own garden or to give away to friends as gifts. Also save a few prunings to make plant supports (see p.22) and pile others in a quiet corner, where garden creatures such as hedgehogs and beetles will make their homes.

These little plant pots are made from a few layers of newspaper, some brown paper packaging and a large leaf from my garden. I made them using a little wooden pot-making press, which I was given as a gift, but you can simply wrap the paper or leaf around a glass and fold the ends under at the bottom. They are biodegradable, so you can plant the pot and seedling together when roots develop.

CHAPTER 2

FEBRUARY

The end of winter is now in sight and, with spring just around the corner, the shoots of perennial plants are beginning to peep up through the soil as the garden stirs into life. My heart soars as the snowdrops make an appearance, together with the first daffodils, gifts for our efforts back in autumn when we planted the bulbs.

Seeds ordered last month have now arrived and I'm ready to start some of them off in my greenhouse or a spare windowsill, excited by the prospect of seeing the first new shoots start to sprout. I also take a few quiet evenings to order some summer bulbs for rewards later in the year.

The light levels are noticeably higher in February, lifting my spirits and boosting my energy levels. I have attention deficit hyperactivity disorder (ADHD) and my mind is always busy, so I like to keep active in the garden, making new beds, cleaning out pots ready for sowing or planting, and pruning my summer-flowering clematis. Try planning your own garden, too, and start the ground work now.

IN SEASON

There are so many plants to enjoy in February, with the show becoming increasingly spectacular as the weeks pass. The hellebores are out in full force now, alongside a medley of early-flowering bulbs sparkling in the winter sun.

WHAT'S IN BLOOM?

The cheapest and easiest ways to create a late winter garden awash with colourful flowers is to buy large bags of dry bulbs and plant them in autumn and early winter (see p.129).

Crocus (*Crocus* species)

One of the first spring flowers to bloom, plant this bulb en masse in autumn in lawns and around trees and shrubs.

Snowdrop (*Galanthus*)

Snowdrops put a smile on my face and remind me of home. I grow two types – the classic white *Galanthus ikariae* and *G.* 'Hippolyta' with its green-edged centres – and bring a small bunch indoors to brighten up my home.

Winter aconite (*Eranthis hyemalis*)

This diminutive beauty, with its little cup-shaped yellow flowers, brings cheer to my winter garden. It's very easy to grow, too, and will spread over time. Like most spring bulbs, it dies down during the summer so take care not to dig it up while planting.

Cyclamen (*Cyclamen coum*)

Another favourite bulb, this little cyclamen produces pink or magenta flowers above kidney-shaped, dark-green leaves, which are often decorated with silver and white splashes.

MINDFUL MOMENT

As a busy mum of two boys, my life can be very hectic and I can forget to stop and take a moment to just 'be'. My way of calming my thoughts is to focus on something in the garden – maybe a beautiful leaf, the first leaves to appear on my seedlings, or a flower. I often watch the world through the lens of my camera and think this makes me more aware of the beauty around me. I take pictures to share with you, and to enjoy and treasure. My heart sings when I see my sons doing the same, knowing I've passed this gift on to them.

SPOTTING WILDLIFE

Frog

When my boys were little, I spent hours walking with them from one pond to the next in my local area, checking to see if the frogs had arrived. Spotting some frogspawn was always a joy and still is, even though we are all older. These amphibians seem to time their courting to coincide with Valentine's Day – finding their way to my pond on that very day and cavorting around for a few weeks, before leaving behind their gift of frogspawn floating on the water. Any small pond will provide them with the perfect love nest, after which they will retreat to damp corners of the garden, where they feast on slugs and snails, helping to keep these unwanted creatures in check.

Clockwise from top left: wood anemones (*Anemone nemorosa*) start to bloom at the end of the month, their starry flowers brightening up shady spots; snowdrops are among my favourite bulbs, marking the end of winter; look out for frogs in ponds in mid-February; crocuses come in a vast range of colours to adorn your lawns and beds.

Clockwise from top left: plant the bulbs of lilies, including Turk's cap types, in a sheltered spot outside in late February; use rose prunings to take hardwood cuttings; I sow some hardy plants from seed this month, raking the soil to form a crumbly soil surface to promote good germination; layered climbers may be ready to repot.

WHAT TO GROW IN FEBRUARY

The end of the season for bare-root plants is fast approaching so order yours now for the cheapest shrubs, trees, roses and climbers. Some may be reduced even further now, as nurseries prepare for spring and their new stock arrives.

SOW INDOORS

Flowers

Hardy annuals can be sown successfully outside in beds in late February (see below), but to protect them from inclement weather, waterlogged soils, and slug and snail attacks, I usually start them off indoors at this time of year. Also sow perennials, such as hollyhocks (*Alcea rosea*), achillea, and hardy geraniums that flower in the first year, which need to get off to an early start (see p.25). Tender plants I sow in February include:

- Snapdragon (*Antirrhinum*)
- Larkspur (*Consolida regalis*)
- Lobelia
- Bedding geraniums (*Pelargonium*)

Herbs, vegetables, and fruit

- Start sowing tomatoes from seed in a warm spot indoors.
- Propagate some herbs by rooting them in water, using inexpensive packets of stems from the supermarket (see also p.116).

SOW OUTDOORS

Flowers

Some hardy plants can be sown outside at the end of this month, providing the ground is neither frozen nor waterlogged. In gardens with heavy clay soil, sow them in pots inside. Try these in beds of free-draining soil:

- Pot marigold (*Calendula*)
- Love-in-a-mist (*Nigella*)
- Wildflower seeds, including poppies and cornflowers
- Pansies and violas (*Viola*)

PLANT OUTSIDE

Sweet peas under cloches

Sweet peas sown in autumn will survive temperatures down to –8°C (18°F) if you plant them outside now, but they need a sheltered position such as next to your house wall, under a cloche, or in a cold greenhouse to protect them from strong winds and heavy snow, which will damage the stems.

Lily bulbs in pots

Get your lilies off to a head start by planting the bulbs in a sheltered spot beside the house. Plant them about 20cm (8in) deep in pots of peat-free compost and set them in a sunny spot when the shoots appear later in spring.

Bare-root plants

As well as shrubs, climbers, and trees (see p.13), you may find a few perennials, including geraniums, rudbeckias, echinaceas, and hostas sold as bare-root plants, and these can also be cheaper than those grown in pots.

PROPAGATE NOW

Continue to take hardwood cuttings

Multiply your shrubs and roses (see p.156).

Layer young bedding plants

Not many people think of layering tender trailing bedding plants such as *Bacopa* or *Lobelia*, which are usually available from garden centres at the end of February. Simply push a flexible non-flowering stem into a separate pot of peat-free compost and pin it in place with wire or a paper clip. Keep the compost moist, and it will then root to produce more plants. This is really rewarding, especially at this time of the year (see also p.26).

MAINTENANCE MUSTS

Leave wildlife habitats intact

Delay cutting back all your perennials and ornamental grasses until the end of the month, as they continue to provide habitats for wildlife that will still be taking shelter in hollow stems and in the soil, while the larvae of moths are nestled in the leaves and grass cuttings that I leave for them over the winter.

Create plants supports

In February and March, I weave all my plant supports using pruned branches, before the plants that need staking start growing, so I don't damage them. Simply insert a few upright stems around your emerging plants and use long flexible stems to weave in between them. These natural supports soon become almost invisible in summer.

Prune fruit bushes

Cut back all the stems of autumn-fruiting raspberries to the ground. Also prune red- and whitecurrants to create an open-centred bush, with eight to ten stems on a short central stem. You can also pop your prunings in a glass or vase of water to root and make a few new plants for free (see p.116).

Prune wisteria

Cut back the shoots you pruned in summer (see p.115) to two or three buds.

Prune summer-flowering clematis

Towards the end of the month, before active growth begins, cut the stems of clematis that flower in July or later in the year down to 30cm (12in) from the ground. Clematis that flower earlier in summer should just be taken down to the first or second set of healthy buds.

Cut back ornamental grasses

At the end of February, just before new growth appears, clip deciduous grasses to within a few centimetres of the ground.

Prune overwintered hardy fuchsias

Take the side-stems of these plants back to one or two buds. In hard winters, the fuchsias may die back completely, but the stems will re-emerge as the weather warms up in spring.

Trim winter-flowering heathers

Remove faded flower stems to prevent your plants from becoming leggy and to promote bushier growth with more flowers next winter.

Remove faded flowers from winter pansies

This prevents them from setting seed and encourages a flush of new flowers when the weather warms up. Alternatively, leave a few flowers to set seed, which you can then harvest and sow for new plants later in the year.

February is a good time to start sowing a range of flowers indoors to decorate the garden in the coming months. Try perennials such as hollyhocks (*Alcea rosea*), achilleas, campanulas, penstemons and salvias, many of which will flower later in the summer, filling your beds with blooms for the cost of a few packets of seed.

1.

2.

3.

4.

ANYA'S PROPAGATION MASTERCLASS

SOWING PERENNIALS FROM SEED

One of the easiest ways to save money is to grow perennial plants from seed. Perennials flower in spring, summer or autumn, after which most will die down over winter before popping up the following spring.

Sowing any plant from seed is usually a money-saving option, but perennials, which can be very expensive when bought as mature plants, offer the best value. You can save even more money by collecting seed from perennials from your or a friend's garden.

However, you may have to be patient to reap the rewards, as perennials can take a while to flower, but I think the journey is often as enjoyable as the destination. Just as you travel on a train watching the world go by, I love watching my seedlings grow and mature. Some perennials will also flower in the first year of sowing, if you start in February.

You will need

Seeds
Upcycled pots or seed trays (see p.14)
Peat-free multipurpose or seed compost
Horticultural grit

1. First, wash out some old food or seed trays and fill them with moist seed compost or peat-free compost mixed with a handful of horticultural grit. Using seed collected from your garden plants or new seed, sow thinly at the depths recommended on the packs. Look up the sowing depths online if you are harvesting your own seed.

2. Set them on a warm windowsill, and cover with an old plastic bag propped up with four sticks set around the pot or seed tray. When the seedlings emerge, remove the plastic bag to prevent damping off disease.

3. When the seedlings have a few sets of leaves, transplant them carefully, holding them by a leaf not the stem, to their own pots of peat-free compost.

4. Keep the plants well watered, and grow on indoors. Plant outside after the frosts in June.

When to sow perennials

Spring or early autumn are the best times to sow perennial seeds.

First-year flowering perennials

Try these plants which will produce flowers a few months after sowing: achillea, agastache, hollyhocks (*Alcea rosea*), daisies (*Bellis perennis*), *Campanula lactiflora,* cornflowers (*Centaurea montana*), chrysanthemums, *Delphinium elatum*, *Echinacea purpurea*, *Erigeron karvinskianus*, sea holly (*Eryngium*), lupins, *Monarda*, gaura (*Oenothera lindheimeri*), poppies (*Papaver*), penstemons, salvias, verbascums, and verbenas.

ANYA'S PROPAGATION MASTERCLASS

LAYERING SHRUBS AND CLIMBERS

While much of the garden is resting, you can start the process of making plants for free by layering shrubs and climbers from the end of February in mild areas and in March in colder spots.

One of the easiest ways of propagating new plants, layering is a great method for beginners to try, since it requires very little equipment and aftercare. Unlike cuttings, which need careful watering to survive, layered shoots form roots while still attached to the parent plant, which has an established root system that keeps them alive. You also end up with a plant genetically identical to its parent, like you would do with a cutting.

To start layering, look for plants with flexible or long stems that you can bend down to soil level or a pot set on the ground.

You will need

Plant with flexible stems
Sharp clean knife
Recycled pot
Peat-free compost
Garden wire

1. Select a strong, healthy, non-flowering shoot at the base of the plant, so it can touch the soil surface without breaking. Then make a cut through the underside of the stem at a leaf joint (remove the leaf first if the plant is in leaf). From my experience, you can miss out this stage if you like, as many plants will still root without cutting the stem.

2. Place the cut edge into a pot of peat-free compost, or dig a shallow trench in the ground and plant it in that. Cover with compost or soil and pin the stem in place with some garden wire. If securing the stem in the ground, use a bamboo cane to prop up the rest of the stem, so that it does not trail on the ground and rot.

3. Press the compost or soil over the stem, and keep the cutting well watered. Then wait for about a year, or until you see new shoots developing from the cutting.

4. Once a good root system has established, you can sever the rooted stem from the parent plant. Keep potting it on into larger containers until you are ready to plant it out.

The best time to layer plants

Early to mid-spring or autumn are the best times to layer your plants.

Suitable plants

Most climbers, including clematis, wisteria, jasmine, honeysuckle (*Lonicera periclymenum*), and *Parthenocissus* species. Shrubs with flexible stems, such as camellias, forsythia, witch hazel (*Hamamelis*) and rhododendrons.

1.

2.

3.

4.

CHAPTER 3

MARCH

Spring offers endless opportunities for every gardener, no matter what size or type of outdoor space you have to work with. Growing something this year will bring you lots of positivity, so buy some seeds and get started, now that the light levels are increasing and those tiny capsules are ready to burst into life. Remember, too, that the process of sowing and taking cuttings from your existing plants is as rewarding as the final results, so take time to enjoy the moment.

In this chapter, I've included information on plants that are particularly rewarding to plant or sow in March. However, before you start, always read the seed packets carefully or check online that the plants you plan to grow are suitable for the type of soil and light conditions in your garden. This will ensure you get the best results for your investment and that you don't waste time or money on plants that won't thrive.

Clockwise from top left: the flowers of *Daphne odora* and its cultivars produce a wonderful scent; sweet violets, *Viola odorata*, have self-seeded all around the garden; *Iris* 'Katharine Hodgkin' is a beautiful blue form of the ground-hugging *Iris reticulata*; primroses brighten up beds, borders, and containers in early spring.

IN SEASON

We all need a little uplifting colour after the winter and your garden will be offering endless sources of positivity now as plants start to wake up. Look around your plot for beautiful blooms and keep an eye out for free seedlings to pot up.

WHAT'S IN BLOOM?

The spring bulbs that you planted in autumn (see p.129) and those that have naturalized over the years will be peeping through the soil now, with the daffodils taking centre stage.

Daffodil (*Narcissus*)

There are so many beautiful daffodils to choose from, but my absolute favourites are the pale-coloured varieties 'Actaea', 'Ice King', and 'Thalia', which look like they are made from silk. I also love the little species *Narcissus bulbocodium*, which is great for pots. Choose a selection that will flower at different times, from February to late April, to extend the season of colour.

Dwarf iris (*Iris reticulata* and *Iris unguicularis*)

When the little *Iris reticulata* start flowering in my garden, I know that spring is on its way. Like tiny jewels shining bright through the wind and rain, they look lovely planted in pots, troughs, and lawns, or in a gravel bed. One of my absolute must-haves, plant the little bulbs in the autumn.

Striped squill (*Puschkinia scilloides* var. *libanotica*)

These diminutive bulbs are an absolute joy when they appear after the snowdrops in March. Their dainty blue-striped white flowers also have a wonderful scent if you get up close to them, and they are happy growing in full sun or dappled light shade under deciduous shrubs or trees, or try them in pots.

Primrose (*Primula*)

Easy to grow, these classic spring flowers are such a great addition to any garden. A couple of my favourite March-flowering varieties are *Primula* 'Gold-laced Group', with its maroon and yellow flowers, and the deep pink *Primula japonica* 'Miller's Crimson', which flowers later in April. Divide them after flowering.

Daphne

An evergreen shrub that's loved for its highly fragrant early spring flowers. While daphne isn't easy to propagate, for cheaper plants, check the reduced section at the garden centre in late spring or early summer, after flowering.

Wallflowers (*Erysimum cheiri*)

I grow these scented biennial plants from seed the previous year, and look forward to their yellow, red, and orange blooms that do a great job of brightening up my beds and containers, alongside my spring bulbs.

Windflower (*Anemone blanda*)

These small daisies grow from stick-like rhizomes planted in autumn, and produce carpets of dainty blue and white flowers in early spring under deciduous trees and shrubs.

Holly-leaved hellebore (*Helleborus argutifolius*)
One of my favourite hellebores, this large perennial has wonderful architectural spiny leaves and pale green cup-shaped flowers that last for many weeks when they unfurl in late winter. It self-seeds and the seedlings come true to type, so remember to harvest them.

Camellia
The flowers of this shade-loving shrub come in a wide range of colours, from white to pink and red, with something for everyone. The only proviso is that it prefers acidic soil – test yours with a soil-testing kit from the garden centre. If yours is alkaline, you can grow one in a pot of ericaceous compost.

Bergenia
An easy-to-grow ground-cover perennial with overwintering leaves, it produces spikes of small flowers in shades of white and pink in spring that provide food for early flying pollinators. Bergenia will thrive in full sun and partial shade in a well-drained soil.

MINDFUL MOMENT

Growing plants and the science of horticulture may seem complicated, but gardens for me simply mean love, legacy, and passion. They have a soft and feminine side, and they can bring both tears and happiness. The natural world can also help to heal broken hearts.

If you feel down, go outside and feel the soil between your fingers, take some cuttings, or sow a few seeds. As soon as cuttings and seedlings start growing roots, you'll see some light and create your own little world of positivity. I look at my garden as a playground for my creativity, not worrying about making everything perfect, but simply using it as a way to engage my mind and enjoy a few peaceful and joyful moments in the fresh air, away from my busy world.

SPOTTING WILDLIFE

Peacock butterfly
We tend to associate butterflies with the summer garden, but the stunning peacock can be seen on warm days in March, emerging from hibernation to feast on the spring flowers. It is named after the distinctive blue and yellow eye-like markings on its black wings, which resemble a peacock's tail feathers, and unlike many butterflies it is not considered to be threatened. You can coax these colourful visitors to your garden with a range of pollen-rich plants that bloom at different times to feed them year round, from mahonias that provide winter fuel to buddleia in the summer months and ivy for autumn feasts.

These butterflies are long-lived, and their whole lifecyle takes under a year. The female lays her eggs on nettles in May, and we have plenty of those in the fields around my garden that act as nurseries for the caterpillars that emerge a couple of weeks later. They spin a silk web around themselves for protection before venturing further afield and pupating in July. The adults then emerge later in July and August, and hibernate over winter in dark spaces such as sheds and holes in trees.

A little known fact about peacock butterflies is that when threatened they make a hissing noise by rubbing their wings together to scare off predators. The startling eye-like markings on their wings also help to dissuade would-be attackers from striking.

It is always exciting to see butterflies in the garden, but I was surprised to see this beautiful peacock butterfly visiting my garden in late March. Unlike some of its cousins, this little beauty hibernates over winter and will come out to feed on mild days in spring, so make sure you have nectar-rich plants to feed them at this time.

Clockwise from top left: the cuttings I took from *Viburnum opulus* 'Roseum' produced early flowers in my warm greenhouse – it usually flowers in April; sow the seeds of *Campanula lactiflora* (top right), *Cerinthe major* (bottom right) and *Echinacea* (bottom left) now for a garden full of flowers later this year or next.

WHAT TO GROW IN MARCH

Sowing and planting begins in earnest now that the light levels and temperatures are increasing, prompting plants to come to life. Seeds sown indoors are more likely to survive but need nurturing, so sprinkle some hardy types directly into beds if you're short of time.

SOW INDOORS

Flowers

Flowers sown from seed in early March will have grown into sturdy plants when it's time to plant hardy types outside later in spring, or after the frosts if they are tender. However, if like me you don't have much space to store seedlings in a greenhouse or on a windowsill, stagger your planting over a few weeks. Try a few of my favourites, which include:

Annuals

- Bishop's flower (*Ammi majus*)
- Snapdragon (*Antirrhinum majus*)
- Pot marigold (*Calendula*)
- Honeywort (*Cerinthe major*)
- Spider flower (*Cleome*)
- Cosmos (see p.42)
- Spanish flag (*Ipomea lobata*)
- Nasturtium (*Tropaeolum majus*)
- Tobacco plant (*Nicotiana*)
- Laceflower (*Orlaya grandiflora*)
- Annual phlox (*Phlox drummondii*)
- Painted sage (*Salvia viridis*)
- Sweet scabious (*Scabiosa atropurpurea*)
- Starflower scabious (*Scabiosa stellata*)
- Statice (*Limonium*)
- Sunflower (*Helianthus annuus*)
- Zinnia

Perennials

- Granny's bonnets (*Aquilegia*)
- *Campanula* species
- Brook thistle (*Cirsium rivulare* 'Atropurpureum')
- Dahlia (seed is cheaper than tubers)
- Coneflower (*Echinacea*)
- Globe thistle (*Echinops*)

Dahlia and begonia tubers

Plant dahlia tubers if you haven't done so already (see p.13). Also plant begonia tubers in pots of peat-free compost, about 2.5cm (1in) deep with the hollow side up. Store in a frost-free place until June.

Herbs and vegetables

- Dill (*Anethum graveolens*)
- Chillies (*Capsicum annuum*)
- Coriander (*Coriandrum sativum*)
- Basil (*Ocimum basilicum*)
- Parsley (*Petroselinum crispum*)
- Tomatoes (*Solanum lycopersicum*)

SOW OUTSIDE

First, check that the soil is not frozen or waterlogged, then sow hardy flower seeds at the depths recommended on individual seed packets, or just scatter them around and enjoy!

Annuals

- Bishop's flower (*Ammi majus*)
- *Bupleurum rotundifolium*
- Cornflower (*Centaurea cyanus*)
- *Clarkia*
- California poppy (*Eschscholzia californica*)
- Stocks (*Matthiola incana*)

Perennials

- Cupid's dart (*Catananche caerulea*)
- Red valerian (*Centranthus ruber*)
- Foxtail barley (*Hordeum jubatum*)
- Hardy geranium
- Wildflower mixes

PROPAGATE NOW

Take softwood cuttings

Increase perennials and shrubs (see p.70).

Take basal cuttings

Lift and pot up the young shoots that grow beside plants such as achilleas, delphiniums, phlox, and lupins.

Divide perennials and bedding plants

When buying bedding such as nemesia, divide up the root ball to make a few plants from one. Also divide large clumps of perennials that are not flowering well (see p.146).

PLANT OUTSIDE

Rooted cuttings

On warmer days I start placing tender young plants that I propagated the previous year, such as Salvia 'Amistad', outside in their pots during the day and bring them in again at night. The extra light helps to boost growth.

Roses

If you took hardwood cuttings of roses the winter before last (see p.156) and they have formed roots, transplant them into their individual pots or plant in the ground now. Also continue to plant bare-root plants.

Strawberries

Buy seedlings online or ask locally if anyone has some runners (baby plants) to offer you. At my allotment, people often leave a tray of free runners just outside the gates.

MAINTENANCE MUSTS

Feed young potted plants

Start feeding potted plants once a week using nitrogen-rich fertilizer. I make my own from stinging nettles and grass cuttings, or you can buy seaweed fertilizer if you prefer, diluting it more than recommended for mature plants. I do it on Fridays and call it 'Feeding Friday' so I don't forget or overfeed.

Mulch your beds

Spread a 5–10cm (2-4in) layer of homemade compost or grass cuttings (only use grass if you do not apply fertilizers containing herbicides) over the soil surface, leaving a gap around shrub and tree stems.

Mow the lawn

Start mowing when the ground is dry and set your mower on its highest setting.

MONEY-SAVING IDEA

Buy packs of rosemary, mint, thyme and oregano from the supermarket, then snip a tiny section from the base of each stem. Pop them in water, set on a windowsill and pot up when roots have formed. This is quicker than growing them from seed and can be cheaper.

Clockwise from top left: take basal cuttings by removing young shoots growing beside a parent plant and potting them up to grow on; divide perennials such as hardy geraniums; propagate stems of vigorous perennials such as *Verbena bonariensis* by rooting them in water; plant rooted cuttings taken the previous summer.

1.
2.
3.
4.

ANYA'S PROPAGATION MASTERCLASS

DIVIDING BULBS

One of the easiest ways to increase your spring flower without buying new bulbs is to divide up established clumps. This prevents plants from becoming congested, which often means they don't flower well, and encourages new clumps to form.

Dividing bulbs couldn't be easier and the new clumps you create are ideal for filling gaps in your beds, borders, and pots that are in need of some spring colour. Snowdrops (*Galanthus*) are also easier to grow from divisions than dry bulbs planted in the autumn.

So, if you have some congested snowdrops, daffodils, grape hyacinths (*Muscari),* or other bulbs, simply follow these easy steps.

You will need

Fork
Spade or trowel

1. Using a garden fork, dig up a clump of bulbs, making sure you take a large section of soil around them to avoid damaging them. You can also lift congested bulbs from a pot.

2. Using your hands, gently tease the bulbs apart. Take care not to damage the bulbs or their fine roots as you do so.

3. Remove any bulbs that look diseased, dead, or shrivelled. Then plant the healthy sections at the same depth as they were previously growing in the soil or container. Make sure their new location suits their site and soil requirements, or they may fail to bloom, and try to space each of the bulbs so that they are close together but not touching one another.

4. Firm the soil around the bulbs with your hands and water well.

Aftercare

After planting, leave the foliage to die back naturally, and your bulbs should then flower the following year.

The best time to divide bulbs

Mid-spring to early summer for spring bulbs.

Suitable plants

Alliums, anemones, glory of the snow (*Chionodoxa*), crocuses, winter aconites (*Eranthus hyemalis*), snowdrops (*Galanthus*), grape hyacinths (*Muscari*), daffodils (*Narcissus),* scillas, and species tulips.

EASY MONEY-SAVING IDEA

If you have areas that would benefit from colourful spring flowers, buy cheap dry bulbs in autumn, or potted-up varieties on sale at the garden centre after they have flowered. Plant where needed for flowers next spring.

PROJECT OF THE MONTH

FREE SEEDLING HUNT

One of the cheapest ways to increase your stock of plants is to go on a seedling hunt in March. Many plants self-seed, usually distributing their wares from little seed cases that form after flowering, treasures that you can retrieve and grow on for free.

Plant seeds that fall on the soil, a gravel driveway, pots of compost, or other hospitable place in the spring, summer, or autumn will germinate immediately, if the temperatures are warm enough, and you can search for them throughout the growing season.

However, seedlings are often easier to spot in spring, when your beds and borders are not as full. They also have more space and light to grow at this time of year, and some seeds may wait until these favourable conditions arrive before germinating, so your free store may also be richer now.

Seedling or weedling?

When hunting for baby plants, you first need to identify them, and ensure that you will be nurturing a treasured plant seedling and not a weed. This becomes easier once the seedling has a few sets of leaves – those that first appear are called seed leaves and often look different from the 'true leaves' that grow after that.

Check plants nearby, from which the seeds may have fallen, to see if you can match the leaves up. Also get to know the common weeds that grow in your garden, so you can also recognize those, too, and remove them.

Planting seedlings

Once you have identified your plant prizes, you can simply dig them up, ensuring you also remove some of the soil surrounding the roots to protect them. Then either replant them immediately in a gap in your bed or border, making sure they have sufficient light and space to grow on to maturity. Alternatively, plant them in pots of peat-free compost or garden soil to grow on, if you are not sure where to put them or wish to grow them in containers. Or leave them where they are, if that's where you would like them to establish. Some may grow unaided in their new positions, but keep checking that surrounding plants don't shade them, which may kill them.

MONEY-SAVING TIP

My iPhone has a free app that identifies plant images for you. Take a photo of the plant you need to identify, then click on the 'i' icon at the bottom of the phone and it will take you to a web page with the plant details. It's not foolproof but identifies most plants.

Here are just a few of the free plants I found scattered around the garden. Clockwise from top left: *Salvia viridis*, which I am moving to my flowerbed; *Cerinthe major*, a beautiful hardy annual; sweet William (*Dianthus barbatus*), a spring-flowering biennial; another *Saliva viridis*, which I pinched out earlier to create a bushier plant.

FOCUS ON COSMOS

Some annual plants are like old friends – they come every summer and bring so much joy. Cosmos is one of these. Quick and easy to germinate, I sow a few different varieties in March or April for a garden filled with daisy-like blooms.

Cosmos is so popular with gardeners because each plant produces lots of flowers and often blooms for months, from early summer until the first frosts. It also makes a great cut flower for the house and bees, butterflies, and moths love it, too, while birds eat the seeds later in the year. It's an annual plant, which means it completes its life cycle, from germination to the production of seeds, within one growing season, and then dies, so you need to sow it each year to enjoy the summer flowers.

How to grow cosmos

Cosmos is half hardy which means it needs protection until the danger of frost has passed. Sow the seeds indoors from March to May, covering them lightly with a sprinkling of sieved compost or grit, which ensures they have the light they need to germinate.

I sow my cosmos in trays and transplant them into their individual pots when they are large enough to handle. I then pinch out the seedlings when they are about 5cm (2in) tall, taking off the top of each stem so they don't get too tall and spindly, and produce more flowers on bushy stems.

After the frosts, plant them out in well-drained soil in full sun or partial shade – if no flowers appear, they may be in too much shade. Alternatively, plant them in pots – some varieties grow up to a metre or more in height so check this before selecting a suitable container. The tall varieties may also need staking, and protect young seedlings from slugs (see tip on p.84).

Aftercare

Water plants until the roots are established – continue to water those in pots every few days in summer. In July, add a tomato feed every week or two, which encourages more flowers to form, and remove the blooms as they fade.

ANYA'S MONEY-SAVING IDEA

Pop the stem that you removed when pinching out into a glass of water to root and make more cosmos plants for free. One seed can produce up to four plants.

FAVOURITE VARIETIES

I love the tall white *Cosmos bipinnatus* 'Sonata White', which grows to 1.2m (4ft) in height, and the equally tall 'Cupcakes and Saucers', with its assortment of flowers in shades ranging from powder pink to cerise, some with frilly-edged petals. Both look great in a bed or border. 'Lemonade' rings in the changes, its pale custard yellow flowers on shorter stems making it a perfect choice for pots or the front of a flowerbed.

Clockwise from top left: I take cuttings of some of my cosmos to root in water; *Cosmos bipinnatus;* 'Picotee' 'is another favourite variety with tall stems of pink-infused white flowers; I always grow a few cosmos to pick for indoor arrangements; 'Candyfloss Pink Sunrise' and 'Sensation Majestic Mix' work well with salvias in a border.

CHAPTER 4

APRIL

April is a very busy month in the garden and I tend to take shortcuts wherever I can to prevent me from feeling overwhelmed, and to save time and money. If there are seeds that can be sown directly into a bed or a pot outside, instead of starting them in trays indoors and transplanting them, I go for the first option. I love feeling like I'm working in harmony with nature and sowing seeds where they will grow feels exactly like that. I think of myself as Nature's helper – scattering the seeds around the garden, then letting the natural processes take over and watching the magic happen when they germinate a few weeks later.

There are lots of other jobs to do this month, too. Pick and choose those that you want to take on from my easy guides outlined in the following pages. If you have just a small space or not much time for pampering your plants, just try a few ideas that you will be able to manage easily and enjoy.

Clockwise from top left: the ornamental cherry, *Prunus* 'Shōgetsu' is decked with pink blossom in April; honesty's purple flowers are followed by pretty white seedheads; this little species *Tulipa humilis* flowers year after year; *Muscari* 'Baby's Breath' is a favourite of mine.

IN SEASON

I'm always excited to see what the garden delivers in April. The bees are busy stocking up on pollen and nectar from the blossom on my fruit trees, while the borders are dancing with colourful tulips, lifting my spirits every time I walk outside.

WHAT'S IN BLOOM?

Many of the plants that decorate our spring gardens are grown from bulbs planted in the autumn (see p.129), which is by far the cheapest and easiest way to grow them.

Muscari

Grape hyacinths, as they are also known, are perfect for pots and edging beds and borders in the garden. However, I find that they don't perform well if left in pots for too long, so after they have flowered, I transplant them into the soil, where they will thrive from year to year. Great choices include the Magic Series of *Muscari aucheri*, available with blue and white flowers, and the beautiful pale blue flowers of *Muscari* 'Baby's Breath', and *Muscari latifolium*, an interesting species with wide leaves and duo-toned flowerheads.

Tulip

Many tulips only perform for one year and can be quite costly, but they are also a real luxury. They are my special treat and every year I buy the sumptuous pink peony look-alike 'Dream Touch' because I love it so much. Sometimes I'm lucky and find it later in November in a sale – it's always worth checking the reduced section of the garden centre or online for bargains later in the season and, remember, your bulbs will still grow and flower if planted up to mid December, or later. Others I grow are the little small-flowered species tulips such as *T. humilis*, *T. sprengeri* and *T. turkestanica*. These offer better value-for-money since they will bloom again reliably if grown in free-draining soil and a sunny or partly shaded site.

There are also a few large-flowered tulip varieties that come up year after year, including those listed below.

- Tulip 'Apeldoorn' – scarlet
- Tulip 'Artist' – orange and green
- Tulip 'Ballerina' – orange
- Tulip 'Bleu Aimable' – pale mauve
- Tulip 'Daydream' – apricot
- Tulip 'Groenland' – pink and green
- Tulip 'Negrita' – dark purple
- Tulip 'Purissima' (White Emperor) – creamy white
- Tulip 'Shirley' – white with purple-edged petals
- Tulip 'Spring Green' – cream and green

Siberian squill (*Scilla siberica*)

It's worth investing in this little blue-flowered bulb for a long-term planting. It will multiply each year and creates a truly spectacular display in both containers and beds and borders, but I find that it does better when planted directly in the ground. Once they have naturalized, lift and divide clumps of bulbs after flowering (see p.39) to spread them around the garden.

Honesty (*Lunaria annua*)
This easy-to-grow annual has two seasons of interest – first, the purple or white flowers appear in April, then, in autumn, papery white oval seedheads develop, helping to decorate the garden throughout winter.

Ornamental cherry tree
These flowering trees are a real treat and there are so many to choose from, but always do some research before deciding on which one will suit your garden conditions. My favourites for small gardens include *Prunus* 'Shōgetsu', also known as the Blushing Bride tree, *Prunus* x *subhirtella* 'Autumnalis', which flowers in winter and spring, and *Prunus serrula* for its amazing mahogany peeling bark.

Lungwort (*Pulmonaria*)
These carpeting perennials produce small pollen-rich flowers in shades of blue, pink, and white above cream-spotted leaves. The flowers of some change from pink to blue as they age, creating a wonderful two-tone effect. The bees adore it, too, and it's very easy to grow.

MINDFUL MOMENT

Spring always makes me happy and the burst of extra light in April as the days lengthen is guaranteed to lift my mood after a cold, dark winter. This glorious month also reminds me that nothing lasts forever. I sometimes think of April on those gloomy days in January and remember that just as the seasons turn and spring follows winter, good times will return.

My garden helps me to find the positive when I'm experiencing challenges in my life, and going outside and sowing seeds or pruning my plants helps me to stay mindful and keep focused on the beauty around me. Simply watching a tulip open its petals when warmed by the sun, and then close them up again as the air cools in the evening, shows that change is a natural process and occurs minute by minute, day by day.

SPOTTING WILDLIFE

Solitary bee
You may be surprised to learn that most of the bees in our gardens do not produce honey or live in hives, nor do they have a queen or colony. Known collectively as solitary bees, as the name suggests, they live alone. There are over 200 species of solitary bee, and they are responsible for pollinating the lion's share of the plants on the planet. It is therefore essential that we protect their food sources and habitats by planting an assortment of pollen- and nectar-rich plants in our gardens to support them.

Solitary bees are also not aggressive – the males generally have no sting while the females will only sting if they feel threatened. Look out for solitary bees emerging in April and May and, later in the season, you may see the females starting to build a nest. Mining bees, which make up about 70 per cent of the solitary bee population, nest in underground burrows, while others use hollows in walls, hedges, or plant stems. The eggs laid by the female will emerge as larvae, eating the nectar-wrapped pollen ball that the mother has provided for them. They then spend the winter as a pupa, before emerging as an adult the following spring. To provide nesting habitats for them, leave plants with hollow stems over winter and twigs and pruned stems in quiet areas of the garden. Never use insecticides.

I often forage for nettles in April and May to make my own free, natural plant fertilizer. Simply gather a pile of nettles, place them in a bucket, and cover with water. Leave the plants to infuse for a few weeks, then strain. Dilute the straw-coloured liquid with water – about 1 part to 10 – before applying it to your plants.

Clockwise from top left: I harvest the seed from strawflowers in autumn, then sow them in early spring to add to my cutting garden; redcurrants can be planted now; when transplanting seedlings, hold them by a leaf, rather than the stem; sow love-in-a-mist seeds every few weeks in spring and summer for a long season of flowers.

WHAT TO GROW IN APRIL

One of the busiest months for sowing and planting, there is never a dull moment in the garden in April. You can sow tender plants and herbs indoors and scatter seeds outside in prepared beds, while filling the gaps with plants taken from cuttings.

SOW INDOORS

Flowers

Try these easy annual flowers in pots indoors now, and if you have a little patience, why not also try sowing perennials from seed now (see p.25)? You can also continue to sow the plants I listed in March (see p. 35).

- Snapdragons (*Antirrhinum*)
- Cosmos
- Sunflowers (*Helianthus annuus*)
- Heliotropes (*Heliotropium arborescens*)
- Statice (*Limonium*)
- Petunias
- Starflower scabious (*Scabiosa stellata*)
- Nasturtium (*Tropaeolum majus*)
- Strawflowers (*Xerochrysum bracteatum*)
- Zinnias

Herbs

A few annual and biennial herbs, including parsley, basil, coriander, and dill, can be sown indoors in pots and trays in April.

SOW OUTDOORS

If you don't have much time, scatter the seed of these beautiful annual and perennial plants where they are to flower, then forget and enjoy when they pop up. It's amazing what you can achieve in 30 seconds in the garden!

- Greater quaking grass (*Briza maxima*)
- Cornflower (*Centaurea cyanus*)
- Honeywort (*Cerinthe major*)
- Cosmos (in mild city gardens, or sow indoors in pots elsewhere)
- Bunny tails (*Lagurus ovatus* 'Bunny Tails')
- Honesty (*Lunaria annua*)
- Night-scented stock (*Matthiola longipetala* subsp. *bicornis*)
- Love-in-a-mist (*Nigella damascena*)
  A top tip is to sow *Nigella* every four weeks between mid-spring to early summer for continuous blooms
- Opium poppies (*Papaver somniferum*)
- Painted sage (*Salvia viridis*)
- Bishop's weed (*Visnaga daucoides* syn. *Ammi visnaga*)

PROPAGATE NOW

Take softwood cuttings

Perennials and shrubs that are shooting now are perfect for taking cuttings (see p.70).

Take offsets

Increase your stocks of succulents such as houseleeks (*Sempervivum*) (see p.54).

Divide bulbs and perennials

Continue to divide plants (see p.39 and p.146).

PLANT OUTSIDE

Cuttings taken earlier

The cuttings of hardy plants that you took last year can be planted directly in the ground, once they have a few leafy stems and a good root system. Alternatively, pot up those with just a little new growth and frost-tender plants such as the blue-flowered *Salvia* 'Amistad' into their individual pots and grow on indoors until the frosts have passed.

Perennials

Hardy perennials grown from seed earlier in the year (see p.25) may also be ready to plant outside now, or pot on tender types, as for cuttings (see above).

Herbs

Perennial and shrubby herbs such as rosemary, mint, thyme, sage, and oregano, taken from cuttings rooted in water and then potted up last year (see p.116) will be ready to plant out.

Shrubs

Propagated shrubs taken earlier from layering, or semi-ripe or hardwood cuttings (see p.26; p.100; p.156) may have rooted by now. First transplant them into their own individual pots and grow them on until they are vigorous, then plant them in the ground.

Fruits

Strawberries only perform for about three years, after which they should be replaced with new plants that you have propagated the previous year in June by removing the runners (see p.36). When the berries start to appear in a few weeks, add a straw mulch around the plants to keep them dry and prevent rotting. The mulch also controls weeds, deters slugs, and helps to retain moisture in the soil.

MAINTENANCE MUSTS

Divide supermarket herbs

Buy one, make five, by dividing the pots of herbs you buy from supermarkets. Gently pull apart the stems, each with some roots attached and plant them in small, individual pots.

Deadhead daffodils

Allow the foliage of daffodils and other spring bulbs to die back naturally – the old leaves act like rechargeable batteries, absorbing energy from the sun to make food that bulks up the bulbs for next year's flowers.

Make your own plant fertilizer

Pack stinging nettles, grass cuttings, or comfrey into a bucket and cover with water then leave for a few weeks to infuse. Comfrey is a little smelly, so I make this in a bucket with a lid and infuse it for six weeks – nettles and grass only take two (see also p.49). Dilute them – 1 part fertilizer to 10 parts water – before use.

Prune winter-stemmed shrubs

Cut back the colourful stems of willow (*Salix*) and dogwoods (*Cornus*) to the ground. The new growth will then provide a beautiful display of young stems next winter.

Water young plants and seedlings

April can sometimes be warm and dry, so remember to keep an eye on seedlings and young plants and give them a drink if the soil or compost is dry.

Feed currant bushes

As soon as the flowers appear on your currant and gooseberry bushes, start applying a high potash fertilizer or a homemade feed (see above) once a week, and water them well during dry spells.

Clockwise from top left: pot on any cuttings taken last year that have good root systems; plant strawberries grown from runners; when buying new plants such as *Heuchera*, see if you can divide the stems to make new plants; plant out *Astrantia* and other perennials.

ANYA'S PROPAGATION MASTERCLASS

TAKING OFFSETS

Offsets are simply baby plantlets that grow alongside the mother, which you can carefully remove and pot on to make new plants for free. Most succulents can be increased in this way.

Nothing could be easier than taking offsets from plants, and you need no experience or special equipment, making it a great method for beginners to try. These little plantlets also need the minimum of aftercare, so they save both money and time. You may see the offsets of succulents referred to as 'pups' and they are genetically identical to the mother plant.

You will need

Plants to propagate
Sharp, clean knife
Large plate
Small pots or trays with drainage holes
Succulent potting compost, or a 2:1 mix of peat-free multipurpose compost and horticultural grit

1. Remove the baby plants growing next to the parent plant, either pulling them away gently with your fingers or using a sharp, clean knife to cut them off. Take care not to damage any tiny roots attached to the base.

2. Leave the offsets on a plate or chopping board for a day or two to dry.

3. Then place the base of each offset in a pot or tray filled with succulent compost or a 2:1 mix of multipurpose compost and horticultural grit. Water gently.

4. Water your succulents about once a fortnight and guard against overwatering, which will cause them to rot – make sure your pots and trays have drainage holes in the base. Then just wait for the offsets to grow new shoots or leaves, which indicates that they have also formed new roots.

New spider plants for free

Spider plants look great in summer outdoor hanging baskets and can be propagated easily from the babies or 'spiders' that appear at the end of their long, arching stems. You will see that each baby has formative roots growing from the base, which you can plant in a pot of compost, while it's still attached to the parent. Once you see new leaves developing on the baby spider, you can sever the stem and grow it on as a new plant in its own pot.

When to take offsets

Any time of the year, as long as the plantlets have an established root system.

Suitable plants

- *Aloe vera*
- *Agave*
- Spider plants (*Chlorophytum comosum*)
- *Echeveria*
- *Sedum*
- Houseleeks (*Sempervivum*)

This tray of houseleeks (*Sempervivum*) and sedums offers a nursery for small offsets, which I harvest every few months to grow in pots of their own. These hardy little succulents are very easy to care for, and can be left outside in a sheltered spot over winter.

PROJECT OF THE MONTH
SOW A FRILL OF FLOWERS

Sow Mexican fleabane (*Erigeron karvinskianus*) along house walls and steps to soften the hard edges with a frill of pretty daisy-like blooms. Flowering from summer to mid-autumn, and even during the winter in mild areas, these little plants offer amazing value for money.

The perfect project for beginners, all you need to decorate your home is a packet of *Erigeron karvinskianus* seeds, which you will find for sale at garden centres or from online nurseries. Alternatively, if you or a friend already have a plant, you can collect some seeds from the flowers' seedheads from late spring to autumn, as they ripen.

You will need

Erigeron karvinskianus seeds
Pots of peat-free compost (for walls)

1. Take a handful of seeds, and sprinkle them into the gaps along a house wall and cracks between your paving slabs and steps. Make sure that the areas where you sow the seeds never become waterlogged, as this plant likes dry conditions and may rot in wet soil.

2. Water gently after sowing. The seedlings should germinate quickly in April, and stems will start to emerge after a couple of weeks.

3. Leave the seedlings to grow on at their own rate, and the plants should start flowering within about three months.

Walls of daisies

If you want to decorate an old brick or stone wall with these colourful daisies, try sowing the seeds into pots filled with a 50:50 mix of sand and peat-free compost. Keep them watered until the seeds germinate and when the seedlings have established a good root system and a few stems of leaves, simply push them gently into the cracks and secure with a few twigs to wedge the young plants in place, if needed. The seedlings will soon take root in the wall and then flower year after year.

Aftercare

Once mature, Mexican fleabane needs no maintenance at all, even during dry periods in summer. However, I do feed them occasionally with my homemade fertilizer (see pp.49 and 52), which helps to promote more flowers to form, although they will bloom without this treatment. In future years, you will find little *Erigeron* seedlings scattered all over the garden. I collect them up and pop them into pots and containers to add colour at no extra cost. I like the way repeating the daisies also links different areas of the garden.

FOCUS ON DAHLIAS

I used to think dahlias were old-fashioned because my granny grew them. Well, that was a long time ago, and I've since grown to adore their beautiful shapes and colours.

Dahlias have enjoyed a spectacular renaissance in recent years and for good reason. I can't think of any other plant that offers such a diverse range of flowers, and there are tall and short varieties for beds, borders, and containers. They are not hardy, however, so be prepared to move them inside over winter.

GROWING DAHLIAS

Dahlias grow from tubers, which are planted in spring, or you can sow some varieties from seed in early spring (see p.25). I grow my dahlias in our cutting garden, where they offer colour outside and a succession of flowers for the house. Creating a space for cut flowers in your garden or on an allotment is a wonderful eco-friendly, money-saving way of filling your home with fresh blooms – turn to p.135 for steps on how I made mine for almost no cost from seeds and cuttings.

In late April or early May, choose a sunny spot and dig a hole in the ground about 12cm (5in) deep. Spread out the fan of finger-like tubers and cover them with the excavated soil. Alternatively, plant them from January to March indoors in a large pot of peat-free multipurpose compost at the same depth. This is a good way of bringing them on earlier in the year, or if you have heavy clay soil that's prone to waterlogging, which may rot the tubers. When growing the tubers in a pot, keep them under cover until all threats of frost have passed, watering them regularly.

Whether you planted your tubers indoors or outside, shoots should appear a few weeks later – cover any grown in the garden with fleece if a late frost is forecast.

Aftercare

When the dahlia shoots appear, pinch out the tips, which helps them to form more robust plants. They will then flower throughout the summer and autumn but remember to remove the dead blooms regularly to encourage more buds to form. Tall varieties may need staking.

New dahlias for free

Once frost blackens the leaves in autumn I always bring my dahlia tubers into a frost-free place and plant them in pots of old compost to guarantee they will survive whatever the weather has in store. It also makes them easier to propagate the following year by division. To do this, in spring, brush the compost off the tubers and cut them into smaller sections, each with an eye, from which new shoots will form. The eyes are small, round bumps or slightly raised areas on the tuber, near the point where it connects to the stem. Plant these in peat-free compost as described above, and wait for your new, free plants to sprout. Remember that dahlias will not grow from a tuber without an eye.

Here is a selection of my favourite dahlias. Clockwise from top left: 'American Dawn', a decorative dahlia with tall stems of apricot flowers with purple tints; the pompons of 'Jowey Frambo' with the cactus-type 'Hillcrest Royal'; the magenta-pink blooms of 'Thomas A. Edison'; and the mininiature cactus flowers of 'Josudi Hercules'.

CHAPTER 5

MAY

The first signs of summer are in the air, and as the temperatures rise, so do my spirits. I love watching the house martins as they fly in from their winter haunts in Africa and start to repair their nests, ready for this year's brood. Like the birds, May for me is full of hope, as the seeds sown in spring and cuttings taken last year start to deliver on their promises.

Have you also noticed the shade of green that illuminates the garden in May? It's lime green and makes a wonderful foil for the purple and white flowers of alliums, honesty, and sweet rocket. Mine are self-seeded and most are not visible until now, when it feels like someone has lifted a curtain and the show has begun.

All the propagated plants are growing very quickly at this time of the year, too, making my days very busy, but I always take time to visit at least one new garden in May for inspiration.

Clockwise from top left: sow the seed of perennial cornflowers in early autumn for flowers the following spring; easy-to-grow in most gardens, the fragrance of lilac flowers fills the air in May; get up close to the flowers of London pride to enjoy their intricate details; *Aquilegia* blooms come in a host of shapes and colours.

IN SEASON

May ushers in a host of beautiful late spring and early summer flowers in my garden. Many tulips are still in bloom, and the pompon flowerheads of the alliums are opening now, too, while the perfume of viburnums and choisyas fills the air.

May can be a tricky 'in between' month, when many spring bulbs are over and roses and summer perennials have yet to bloom, but if you plan ahead, you can fill the gaps with beautiful plants. As a money-saving gardener, I always prepare my May displays the previous autumn, sowing, propagating, and planting, rather than buying expensive plants now.

Lilac (*Syringa vulgaris*)
I know summer is on its way when my lilac starts to flower in May, filling the garden with sweet perfume. These shrubs can grow quite large, so give yours plenty of space to perform.

Granny's bonnets (*Aquilegia vulgaris*)
One of my favourite spring flowers, my mother grew *Aquilegia* in her garden and I love growing it in mine as a reminder of my childhood. It's easy to grow from seed and will then self-seed around the garden, although the babies may not be the same colour or flower type as the parents, so it's a lovely surprise when different plants pop up.

Sweet rocket (*Hesperis matronalis*)
Another wonderful self-seeder, sweet rocket seems to appear overnight in the gaps in my garden, its pale lilac or white flowers scenting the air in May, drawing me and the bees to its pretty blooms. The flowers are also edible.

Japanese spiraea (*Spiraea japonica*)
These late-spring flowering shrubs come in an array of pinks and whites, and a few such as 'Goldflame' and 'Magic Carpet', also have colourful foliage. My 'Goldflame' was taken from a cutting and I prune it after flowering, removing any green foliage, so that it retains its bright leaves.

Perennial cornflower (*Centaurea montana*)
One of my favourite wildlife plants, the beautiful sculptural flowers of this easy-going perennial will attract all kinds of pollinators. It copes well in dry shade, blooms for many weeks if you deadhead it, and offers great value for money when grown from seed.

London pride (*Saxifraga* x *urbium*)
My heart sings when the little flowers of this ground-hugging succulent appear in May. They look like pink clouds from a distance and the detail on each tiny bloom is incredibly beautiful. Use them as cut flowers, and divide clumps in early summer after flowering.

Spurge (*Euphorbia griffithii* 'Fireglow')
One of my favourite spurges, this one, with its bright orange flowers, appeared in my garden from nowhere. I suspect the seed was in the soil from plants I took from my mother-in-law's garden and it's a lovely memory of her.

Peonies (*Paeonia*)

The stars of the May garden, these drama queens outshine everything else in May. Herbaceous peonies (*Paeonia lactiflora*) have a relatively short season of interest, but they are still worth growing, and my favourites include the creamy-white 'Duchesse de Nemours', cerise 'Miss Eckhardt', and 'Festiva Maxima' with its white blooms with red flecks. Peonies can be divided every 5-7 years in autumn, after which I usually grow the divisions in pots for a couple of years to add to my spring patio displays instead of buying bedding.

7 reasons why peonies may not bloom

If your peonies fail to flower, check the following points to discover why they aren't performing as they should.

1. **Not enough sun** – peonies need a bright sunlit position.
2. **Overfeeding** – a nitrogen-rich fertilizer may result in lush foliage but no flowers.
3. **Late-flowering variety** – check the label to see if yours flowers later in the season.
4. **Plants are too young** – peonies need to be a few years old before they will flower.
5. **The stems are buried** – planting too deeply may result in few or no blooms. The tuberous roots should be no more than 2.5cm (1in) below the surface.
6. **Late frosts** – these can damage the flower buds which then fail to open.
7. **Fungal disease** – the *Botrytis* fungus may have attacked the flower buds, which then fail to develop. Remove spotted leaves and improve air circulation around the plant to prevent fungal disease.

MINDFUL MOMENT

Late spring always makes me feel grateful for the natural world around me. The dawn chorus is at its loudest now, and I love waking to the sound of birds chattering in the trees. When I walk around the garden, picking a few blooms for the house, I'm listening out for chicks in their nests as the parents come in to feed them, insects buzzing around the plants, and the sound of wind rustling the leaves. I'm hypersensitive to sounds, scent, textures, and beauty – it's one of the common traits of ADHD – but Nature also makes me feel calm. Try it yourself. Let your senses enjoy the sights and sound around you: notice the details on the petals and leaves, take in the scents, and explore plant textures with your fingertips to feel connected with Nature.

SPOTTING WILDLIFE

Ground beetles

The ground beetle is an incredible ally in the garden, helping to keep slugs at bay, which may be particularly destructive at this time of year. These little armoured creatures are known as 'generalist feeders' because they consume a wide variety of flora and fauna, including weed seeds, aphids, fly eggs and larvae, as well as slugs. Decomposer ground beetles such as the rose chafer beetle consume rotting wood and break down organic matter, helping to feed the soil and plant roots.

I make habitats for ground beetles by leaving layers of chopped stems on the ground over winter and making wood piles from prunings. They also like open-topped compost heaps that they can access easily, and plenty of plants to provide cover from predators.

Clockwise from top left: ground beetles will help to keep your plants free of pests such as slugs and aphids; spend a few minutes outside each day in spring taking in the colours, scents, and sounds; grow peony cuttings in a pot instead of spring bedding; sow sweet rocket seed in the autumn for scented flowers in late spring.

I plant little *Allium sphaerocephalon* (see p.75) bulbs in the autumn in cell trays, and then leave them in a sheltered spot outside over winter. The shoots appear in spring, when I plant them into any gaps in the border. The bulbs are so small that I risk digging them up if I plant them straight into my borders in autumn.

WHAT TO GROW IN MAY

Late spring is a great time to sow directly into your beds and borders, now that most of the frosts are over, but do keep an eye on the forecast and cover tender plants on cold nights. Self-sown plants can also be potted up and grown on now.

SOW INDOORS

Flowers

You can continue to sow the annual and perennial plants listed in March and April (see p.35 and p.51). Sowing some in small batches every few weeks can extend the flowering season later in the year, too. Also try these other perennials, which will germinate quickly in the warm, light conditions.

- *Achillea*
- *Agastache*
- *Alchemilla mollis*
- Chrysanthemums
- Perennial poppies (*Papaver* species)
- Fringe cups (*Tellima grandiflora*)
- *Veronicastrum*
- Pansies and violas (*Viola*)

Herbs

Continue to sow batches of the annual herbs listed for April (see p.51) to extend the harvest.

SOW OUTDOORS

May is a good time to sow hardy annual flowers directly into beds and to think ahead and sow the biennials listed here, which are plants that germinate and put on growth in their first year, then flower, set seed, and die in the second year. Biennials include some lovely spring flowers such as forget-me-nots, sweet Williams, and wallflowers.

- Hollyhock (*Alcea rosea*)
- Sweet Williams (*Dianthus barbatus*); 'Hollandia Purple Crown' is my favourite.
- Fringed pink (*Dianthus superbus*)
- Foxgloves (*Digitalis*)
- Common teasel (*Dipsacus fullonum*)
- Wallflowers (*Erysimum cheiri*)
- Sweet rocket (*Hesperis matronalis*)
- Honesty (*Lunaria annua*)
- Forget- me-nots (*Myosotis sylvatica*)

PROPAGATE NOW

Take heel cuttings from shrubs

Remove shoots from a perennial or shrub with a little 'heel' (small tail) of bark at the base, which contains high levels of hormones that encourage root growth (see also p.100).

Take basal cuttings

Take these from dahlias, lupins, delphiniums and other perennial plants (see p.36).

Take softwood cuttings

These can be taken from many plants in May from the new growth (see p.70).

Divide perennials

Clumps of pulmonarias, primulas, and other early-flowering perennials can be divided now (see p.146) after they have flowered. Also divide summer-flowering perennials such as hardy geraniums, *Nepeta*, achilleas, and *Echinacea purpurea* before they bloom.

PLANT OUTSIDE

Cuttings taken earlier

Continue to plant out rooted cuttings that you took last summer and autumn.

Annuals

After the frosts in May, all annuals, including tender types, can be planted out now.

Perennials

Continue to plant outside perennials that you have grown from seed (see p.25), including tender types after all threats of frost have passed. Also plant those you have repotted in April, if they now have an established root system and plenty of top growth.

Herbs

Sage, thyme, and rosemary, grown from cuttings taken the previous year, as well as annuals sown earlier in the year, may be ready to plant out in late May.

Fruits

Continue to plant strawberries, if you didn't get round it last month (see p.52). Other fruits, such as gooseberries and currants, can be planted now, but they will only be available as potted plants and may cost a little more than buying them as bare-root plants in winter. However, you can buy a plant and take semi-ripe cuttings (see p.100) or root some stems in water (see p.116) to increase stocks.

MAINTENANCE MUSTS

Pinch out plants

If softwood cuttings taken earlier are growing well, pinch out the shoot tips to create bushier and stronger plants. Also pinch out sweet Williams (*Dianthus barbatus*) before flowering.

Prune spring-flowering shrubs

Plants such as forsythia, kerria, Japanese quince (*Chaenomeles*), *Choisya*, and flowering currants (*Ribes*) can be pruned after they have flowered. Take out dead, diseased, and crossing stems, and then prune to create a balanced shape. Also remove any old wood that is not flowering well.

Deadhead tulips

After the flowers have faded, remove the seedheads to prevent the plants using their energy to make seed rather than bulking up the bulbs that carry next year's flowers.

Try the Chelsea chop

Cutting the stems of perennials such as sedums, asters, and phlox back by half now will result in bushier, more robust plants and help to prevent tall stems flopping later in the year. By doing this to 50 per cent of the stems, you will also increase the flowering season.

Pinch out the flowers on new strawberries

Removing the flowers of new plants allows them to put all their energies into leafy growth and a strong root system, which will give you a better crop the following year.

Continue to water young plants

Any plants in pots and seedlings that you have planted outside in beds will need regular watering as temperatures increase, until their root systems are fully established.

Clockwise from top left: pop the stems you pinch out from young plants into water to root for even more plants for free; sow fringe cups in May to enjoy their tall flower spires next spring; hardy geraniums (right) and achilleas (left) are easy to grow from seed.

ANYA'S PROPAGATION MASTERCLASS

SOFTWOOD CUTTINGS

These cuttings offer a simple way to propagate many perennials and deciduous shrubs, as well as some trees, from the soft, young shoot tips, which root easily but need a little careful nurturing.

There are a few different types of softwood cuttings, and you may see them referred to as nodal cuttings, but the technique is more or less the same for each one. A node is the area on a stem from which a shoot will appear and it can be a little bump or mark, or the point from which a leaf is growing (leaf node). This is where there is a concentration of hormones, from which roots will grow.

You will need

Clean, sharp knife or secateurs
Small pots
Stick
Peat-free cuttings compost or a 3:1 mix of peat-free multipurpose compost and horticultural grit

1. Take a non-flowering shoot about 20cm (8in) long from the tip of a healthy looking stem, cutting just below a node (see above), leaf, or bud.

2. Remove the lower leaves and, if the plant has large foliage, cut the remaining leaves in half – this helps the plant to preserve water and prevents the cutting from wilting.

3. Fill a pot with peat-free cuttings compost, and use a small stick to make a hole in it, close to the edge. Insert the cutting and firm it in gently with your fingers. Take a few cuttings in the same way and insert in the same pot, making sure the leaves are not touching. Then water in gently. Keep the cuttings moist but not wet in a frost-free place indoors.

4. After a few months, you will notice new shoots appearing from each of the cuttings. Carefully lift them out of their pots and repot into a container of their own. When each has a mature root ball, plant outside (when the frosts have passed, in the case of tender plants such as dahlias).

When to take softwood cuttings

Spring and early summer, when new shoots are appearing on plants, are the best times to take softwood cuttings. Pot them up into individual containers by midsummer, and they should survive the winter.

Suitable plants

I use this propagation technique for many plants in my garden, including:

- Hardy perennials
- Tender perennials, such as dahlias and pelargoniums
- Shrubs, including fuchsias, lavender, hydrangeas, and viburnums

1.
2.
3.
4.

PROJECT OF THE MONTH

MIXED HERBS IN HANGING BASKETS

There are so many exciting herbs suitable for a basket and you can opt for old favourites such as thyme, oregano, and chives, or some more unusual varieties that you could grow from seed.

I'm using a hanging basket lined with a jute liner, which absorbs water well. However, all hanging baskets need regular care and watering every day or two in summer, so consider installing an automatic watering system if you are short of time.

Remember that just like many other plants, different herbs have different life cycles. Some are perennials and shrubs and will last for years, while others are annuals or biennials (see p.67).

You will need

Hanging basket and liner
Peat-free multipurpose compost
Scissors and old compost bag
Selection of small herb plants. Here, I've used trailing *Satureja douglasii* 'Indian Mint', variegated mint (*Mentha suaveolens* 'Variegata'), chives, and thyme.

1. I line my baskets with repurposed plastic, which helps them to hold water for longer and keeps the plants thriving. Cut out a circle from an old compost bag and snip a few holes in it, then use it to line the basket.

2. Add peat-free multipurpose compost, so that the basket is about three-quarters full. Then plant the tallest herb in the centre or at the back and the trailing mint and shorter herbs around the sides. Keep the mint in its pot, as it is prone to swamp its neighbours.

3. Leave a gap of about 2.5cm (1in) between the surface of the compost and top of the basket to allow water to collect and filter down to the roots when watering.

4. Secure your basket to a sturdy bracket on the wall close to your kitchen so that you can access the herbs easily while cooking.

Aftercare

Water your basket every day or two from late spring to early autumn. Harvest only a few leaves from each plant at one time, so that they will continue to grow. Replant the herbs in the ground or larger pots after a couple of seasons, as they will soon outgrow the basket. Keep the mint in a pot of its own.

Suitable plants

Annual herbs suitable for baskets include basil, coriander, and dill; caraway, chervil, and parsley are biennials; and mint, chives, and tarragon are perennials; and rosemary, sage, and thyme are shrubs.

1.
2.
3.
4.

Clockwise from top left: *Allium hollandicum* 'Purple Sensation' produces vibrant purple flowers on tall stems in early May.; the white *Allium stipitatum* 'Mount Everest' make a lovely contrast to purple varieties; little *Allium sphaerocephalon* appears later in summer alongside the taller mauve *Allium stipitatum* 'Summer Drummer'.

FOCUS ON ALLIUMS

Just as the tulips are going over, alliums emerge to fill the gap between spring and summer. There are lots of different types to choose from, and I don't know anyone who isn't charmed by their simple beauty.

The classic purple pompon flowers on tall sturdy stems are among my favourite alliums but there are many others, including those with smaller starry blooms that look like little fireworks such as *Allium neapolitanum* Cowanii Group, and *Allium sphaerocephalon*, which produces slender stems of small, egg-shaped dark red flowerheads later in summer. All alliums are easy to grow and ideal for beginners, while the flowers are loved by bees and other pollinators.

How to grow alliums

You can plant dry allium bulbs in the autumn, which is by far the easiest and cheapest way to grow them, or buy potted plants now. This is a more expensive option but will provide immediate colour if you need it, and the flowers will reappear the following year.

Like most bulbs, they prefer a free-draining soil and a spot in full sun or part shade. Plant the bulbs at a depth of about four times the height of the bulb. If your soil is prone to waterlogging, plant them in large pots of peat-free compost and keep them in a sheltered area. Once established, most alliums will form loose groups that come back year after year without replanting new bulbs.

To add alliums to a wildflower meadow or in between other plants in a border, try planting them in trays or pots in the autumn and then transplant into the ground in spring. I find this useful for summer alliums that I may dig up inadvertently earlier in the season when planting my beds.

Aftercare

The flowers produce decorative seedheads, but the stems may start to flop after a month or two, at which point I remove them. Aside from that task, you can leave them be for a show of flowers the following year.

Favourite varieties

I love all the alliums but the following are among my favourites.

Allium cristophii

This short variety grows to about 60cm (2ft) in height and flowers in late May or June, its huge globes of starry violet flowers evolving into dramatic giant seedheads that make fantastic decorations for the house and garden.

Allium 'Globemaster'

Similar in colour and height to 'Purple Sensation' (see opposite) but with larger, more dramatic flowerheads in June or early July.

Allium 'Gladiator'

The impressive large violet heads of this gorgeous allium are sure to make an impact when they appear in May on tall, sturdy 120cm (4ft) stems.

CHAPTER 6

JUNE

With the arrival of summer, I like to take a break to enjoy the show I've created with my seeds and cuttings. Using all my senses, I look at the colours and breathe in the fragrance as I walk around the garden and observe. It's important to do this for two reasons. One, of course, is to simply enjoy the fruits of my labours, but, as a money-conscious gardener, it also allows me to assess what has worked best and to make notes for the future.

Plants must really work hard to be in my garden and by watching and writing down what has thrived and is looking good helps me to focus on my successes and think about why some plants may have failed. I also take lots of pictures to look through over the winter months, before I compile my 'to-do' list for next year.

Although it's good to take some time off, I can't sit still for long, and while I wander around my garden, I'm always on the look-out for shoots to propagate and save money.

Clockwise from top left: the dark purple flowers of *Clematis* 'Étoile Violette' bloom for many weeks in summer; astilbes' feathery flowers make a beautiful textural feature; *Cercis canadensis* 'Ruby Falls' creates a dark foil for my blue geraniums; the flowering dogwood *Cornus florida* is covered with ivory flower-like bracts in June.

IN SEASON

The garden is awash with some of my favourite flowers in June. The roses are in bloom and my large mixed border is packed with early summer perennials, leafy shrubs and clematis which together are creating a riot of colour.

WHAT'S IN BLOOM?

June is often described as the height of the garden flowering season, when long hours of sunlight fuel plants to grow rapidly and put on a stunning performance. Here are a few of my favourites at this time of year.

Masterwort (*Astrantia*)
A good choice for partly shaded spots, this pretty perennial will start flowering in late spring or early summer, its show of unusual pincushion-shaped blooms lasting for many weeks. The flowers come in shades of maroon, pink, or white, and I've grown all of mine from seed or divisions (see p.25 and p.146).

False goat's beard (*Astilbe*)
If you have a damp spot in your garden, fill it with astilbes, which thrive in a consistently moist soil. The plumes of fluffy flowers come in shades of red, pink, and white, and the seedheads offer winter interest, too. One of my favourites is the pale pink *Astilbe* x *rosea* 'Peach Blossom'.

Campanula (*Campanula*)
I grow a few different campanulas, and use the taller milky bellflowers (*Campanula lactiflora*) in my cutting garden. These come in a range of colours, including blue, pink, and white, and – their straight stems look great in a vase. For shadier areas, try the blue-flowered trailing type, *Campanula poscharskyana*, which will grow under shrubs and create a flowery carpet in the cracks in paving.

Eastern redbud (*Cercis canadensis*)
This wonderful deciduous shrub is a must for all gardens. The most popular cultivar is 'Forest Pansy', which produces heart-shaped leaves that are a rich deep red-purple from spring to summer and then turn orange, bronze, and red-purple in autumn before falling. I grow 'Ruby Falls', with dark purple leaves and 'Hearts of Gold' with yellow foliage. Air layer the stems for new free plants (see p.132).

***Clematis* 'Étoile Violette'**
With its deep purple summer flowers, this clematis is one of my favourites for summer colour. I have it climbing over my garage and have also layered it (see p.26) so I now have a few plants twining through the borders.

Foxgloves (*Digitalis purpurea*)
These biennial plants grow leaves in the first year, and flower and set seed in the second, bringing partially shaded areas of the garden to life with their spikes of funnel-shaped pink, peach, or white flowers. Most of my foxgloves are self-seeded, but I love adding new varieties each year, sowing them in trays and planting in the garden in spring. I also collect my own seed and scatter it in all the borders.

Hardy geranium (*Geranium*)
Also known as cranesbills due to the shape of their seedheads, this useful group of perennial plants are so easy to grow, either from seed or divisions, and they flower for most of the summer. Some are tall, growing up to a metre (3ft) in height, while others hug the ground. Many are also happy in some shade. My favourites are the blue-flowered ROZANNE and the pink, nectar-rich *Geranium macrorrhizum* 'Ingwersen's Variety'.

Flowering dogwood (*Cornus florida*)
I love this small tree when it's adorned with small, green flowers surrounded by eye-catching petal-like white or pink bracts. The green leaves also turn red and purple in autumn, and you can propagate it by air layering (see p.132).

Lady's mantle (*Alchemilla mollis*)
Alchemilla looks its best in early June when the frothy lime-green flowers cover the pretty lobed leaves. I mix it with blue, purple, and white blooms and use the flowers as fillers for my floral arrangements.

MINDFUL MOMENT

Every year, I sow my sweet peas from seed in the autumn and build an obelisk in spring, using old stems, to support these beautiful twining plants. The seeds spring to life a few weeks after sowing, just as winter is setting in, and they then sit out the cold months close to the house wall for extra protection, waiting for spring to arrive, before returning to growth.

Now, in June, they are in flower and giving me such joy. This is a great reminder that it's not always the final result that gives us the greatest sense of achievement – the journey there is just as enjoyable. Over the years, I've learned to love the process of growing plants and found success feels even better when I've worked for it. Caring for those seedlings through the darkest months and seeing them shoot up as the light increases helps me really appreciate the beautiful scented flowers I now have in a little vase on my desk.

SPOTTING WILDLIFE

Privet hawk-moth
It's such a treat when we find the spectacular privet hawk-moth in our garden in June and July. The prima donna of the moth world, it's one of the UK's largest hawk-moths and sports an eye-catching pink and black striped body and hind wings. Emerging at dusk to visit night-scented flowers, you may also spot one resting on a tree trunk or fence post during the day.

These beautiful moths are named after the caterpillars, which are bright green with pink and white stripes and feed on privet hedges during the summer, before they descend to soil level to pupate in burrows underground later in the season. Other plants to include in your garden to feed the caterpillars include lilac, holly, viburnum, and forsythia. The moths emerge from their pupae in early summer, when you can see the adults on the wing, feeding on pollen-rich flowers such as honeysuckle (*Lonicera periclymenum*), evening primrose (*Oenothera biennis*), tobacco plants (*Nicotiana*), and jasmine (*Jasminum*).

Right *Cercis canadensis* 'Hearts of Gold', with astrantias, salvias, hardy geraniums, and *Nepeta*, mostly grown from cuttings and divisions.

Clockwise from top left: plant out herbs sown from seed or grown from cuttings; continue to take softwood cuttings from perennials throughout June; sow a few love-in-a-mist (*Nigella*) every few weeks to keep the show going through the summer; I grow dittany (*Dictamnus*) for its tall spikes of fragrant pink flowers.

WHAT TO GROW IN JUNE

Plants are growing at full speed now, and you can sow both hardy and half-hardy annuals directly in the ground to flower later in the summer and autumn. Think ahead, too, and sow some plants for winter and early spring container displays.

SOW INDOORS

Flowers

While the temperatures are often warm enough to sow most plant seed outside in June, slugs and snails can munch them to the ground the minute they germinate, so I like to sow some under cover to give them a little more protection while they mature. Try these for flowers later in the autumn or next spring:

- Winter-flowering violas and pansies – I love the little dark *Viola* 'Back to Black'
- Sweet Williams (*Dianthus barbatus*)
- Wallflowers – (*Erysimum cheiri*) try the red 'Fire King', purple 'Sugar Rush' and yellow 'Golden Jubilee', to flower next spring

Herbs

I continue to sow a few annual herbs such as basil and coriander to extend the harvest.

SOW OUTSIDE

I like to scatter seeds outside in prepared beds in June, where they will germinate later than those I sowed earlier in pots indoors and help to extend the flowering season up to the frosts. Try sowing a few annuals and perennials in this way – the following are robust and the seedlings less prone to slug attacks.

Flowers

- Rockcress (*Arabis*) – sprinkle seed into cracks in paving
- Pot marigold (*Calendula*)
- Canterbury bells (*Campanula*)
- Dittany (*Dictamnus fraxinella*)
- Mexican fleabane (*Erigeron karvinskianus*) (see p.57)
- Fringe cups (*Tellima grandiflora*)

PROPAGATE NOW

Heel and basal cuttings

Continue to take these from dahlias, lupins, and delphiniums (see p.100 and p.36).

Softwood cuttings

Continue to take these from the new growth of perennial plants and shrubs (see p.70).

Rooting in water

I also take stems from *Verbena bonariensis* and other perennials in June and pop them into jars of water to root (see p.116).

Internodal cuttings

Use this propagation method to increase your climbers now (see p.86).

PLANT NOW

Cuttings taken earlier

Most plants grown from cuttings taken the previous year that have a strong root system can be planted out now.

Annuals

You can now plant outside all the annuals you grew under cover earlier in the year, including any tender types.

Perennials

Continue to plant out perennials grown from seed under cover (see p.25), including tender plants, now that the frosts are over.

Herbs

Herbs grown from seed or cuttings, including chives, basil, coriander, and dill, can be planted in their final positions in June.

Fruits

Plant late-season or everbearing strawberries, if you haven't done so already, which will fruit from midsummer. It is a little late to plant out other fruits, such as berries, but you may find some that have been raised in pots are a little cheaper now as nurseries try to clear their stocks. Plant them immediately, and water frequently until they are established.

MAINTENANCE MUSTS

Remove unwanted plants

Dig out pernicious weeds such as ground elder and brambles that may affect the growth of your ornamental plants. Ideally remove the whole root system, and keep pulling out any new growth that reappears. I also hoe off any smaller annual weeds on a dry, sunny day and leave them on the soil surface to shrivel up.

Pinch out sideshoots of tomatoes

Remove the shoots from cordon tomatoes that are growing in the joints between the main stems and side-stems. This prevents the plant putting its energy into leafy growth rather than the fruits.

Deadhead daily

Snipping off or pinching out the dead heads of your flowers will prompt many perennial and annual plants to produce more blooms, so it is worth doing this regularly to extend the display. I do it every day or two as I take a wander through the garden.

Keep slugs at bay

An effective way to keep these molluscs off your young plants is to surround them with a ring of lawn clippings about 2.5cm (1in) deep, which they will eat instead of your ornamentals. Replenish it regularly. Sharp grit also helps if you don't have a lawn.

Feed flowers and fruits

Any flowers that have been in pots for a few weeks will need topping up with a natural fertilizer such as liquid seaweed or homemade comfrey or nettle feed (see p.52). Fruits such as berries will also need feeding every week or two throughout the growing season, when the flowers appear.

Tie in climbers

Keep young climbers such as clematis and roses tied into their supports. Once established, most clematis will then cling of its own accord.

Support new plants

I have a few small supports that I use to prop up young plants so that they don't flop over their neighbours as they grow.

Clockwise from top left: feed flowers grown in containers every few weeks; add plant supports to young plants to prop up the stems; tie in young clematis growth until it starts to cling of its own accord; deadhead *Verbascum* stems to prompt the plant to produce more.

ANYA'S PROPAGATION MASTERCLASS

INTERNODAL CUTTINGS

This propagation method is a good way to increase your stocks of climbers such as clematis and jasmine. Look for stems that are firm and take a few cuttings from each one.

Just to recap, a node is the point on a stem where a side-stem or new leaf will grow from, and, unlike softwood cuttings (see p.70), in this method, you make the bottom cut between two nodes, rather then under one. For success, you need a stem that is fairly well established so reject any soft tips. Then proceed as follows:

You will need

Sharp knife
Firm stems to cut
Recycled pots
Peat-free cuttings compost or peat-free multipurpose compost with horticultural grit
Reused plastic bag and small prunings

1. Cut a stem just above a pair of leaves at the top of the cutting. Then make the bottom cut halfway between two leaf nodes, as shown.

2. Remove about one-third of the foliage, so you are left with just a few leaves to sustain the cutting while the roots grow.

3. Fill a pot with peat-free cuttings compost or multipurpose mixed with a little horticultural grit. Then gently push the cutting into the compost, almost up to the top node, setting it at the edge of the pot.

4. Repeat with a few more cuttings to fill the pot. Water the pot and store outside in a bright place, but out of direct sun.

TOP PROPAGATION TIP

When taking cuttings of large-leaved plants in summer, place a plastic bag supported by four sticks over each pot. This traps moisture and prevents the stems from wilting in the heat. Taking the bag off now and again to increase the airflow around your cuttings will help to prevent fungal diseases. However, don't add a plastic bag over small- or silver-leaved Mediterranean plants such as lavender and sage, which are prone to rotting in the humid conditions.

When to take internodal cuttings

From early to late summer are the best times for internodal cuttings; they should then have rooted before winter sets in.

Suitable plants

- Clematis
- Cobaea
- Jasmine (*Jasminum*)

1.
2.
3.
4.

PROJECT OF THE MONTH

MAKE A FLOWERY SEATING AREA

Wooden chairs that you no longer have use for in the house can be transformed into beautiful outdoor furniture with a few licks of exterior paint, or buy pre-loved seats to upcycle.

I found a couple of old wooden chairs for free on Facebook market place, which I've used to expand the seating areas in the garden, so that I can enjoy it from different angles. It took just one afternoon to revamp them and they now look as good as new, while some of the plant pots I've used to decorate the area are also second-hand, keeping the cost to a minimum.

You can also find pre-loved furniture and containers on eBay, local Facebook groups, or your nearest recycling centre. Alternatively, look around your home or ask friends if they have tables and chairs they no longer need.

You will need

Wooden chairs to upcycle
Fine sandpaper
Stiff brush
Exterior wood paint and paintbrush
Selection of plants
Pre-loved pots
Peat-free compost
Horticultural grit

1. Sand down the chairs, and use a stiff brush to remove any sawdust. Then paint them with an exterior wood paint in a colour to suit your scheme. I opted for a marine blue, which works well in my garden setting.

2. Place your chairs on a hard surface such as paving or gravel – the legs may rot if placed on soil and a lawn, where the grass will also die underneath them.

3. I like to get up close to my plants to see their details and enjoy the colours and scents, so I have surrounded my seats with pots of flowers, grasses, and succulents. Try Mexican fleabane (*Erigeron karvinskianus*), lavender, *Nepeta* × *faassenii* 'Purrsian Blue', and the airy looking feather grass, *Stipa lessingiana*, in large pre-loved pots of peat-free compost. To save money I sometimes mix new compost with old, or even a little garden soil.

4. Plant low growing plants such as Echeverias, houseleeks (*Sempervivum*), which I increase from offsets (see p.54), in shallow pots of compost mixed with a handful or two of horticultural grit. You can then gaze down at them while you relax in your chairs.

Aftercare

Keep all the plants in pots well-watered from spring to early autumn, and repaint the chairs every couple of years to keep them looking good and to prevent them rotting.

Clockwise from top left: the pink shrub rose 'Olivia' has a beautiful fruity fragrance; the rambler 'Crimson Shower', grown from a cutting, decorates the front of my house; 'Generous Gardener' is another favourite, with its climbing stems of scented pale pink flowers; the stems of 'Adélaïde d'Orléans' ramble through my apple tree.

FOCUS ON ROSES

Roses are at the top of my list of favourite plants and I love them for their colourful flowers and fragrance. Propagating roses can also be really meaningful, allowing you to pass down cherished plants from generation to generation.

There are many types of rose to choose from, ranging from the elegant hybrid tea roses with their large, classic flowers, to bushy shrub roses and ground-hugging dwarf types. There are also tiny roses for growing in containers, and lofty ramblers that will scale a tall tree, decking the boughs with clusters of small flowers in early summer. Once established, you can also propagate your plants to produce more wonderful blooms, too.

HOW TO GROW ROSES

When buying a new rose, it's best to wait until late autumn when bare-root plants become available, as these will be cheaper to buy than those grown in a pot and sold at other times of the year. However, June is the perfect time to choose a rose, since they are in bloom now and you can see exactly what they will look like.

Bare-root roses must be planted soon after they arrive, as long as the soil is neither frozen nor waterlogged – you can plant them temporarily in a pot of peat-free multipurpose compost if the conditions are unfavourable. Containerized roses can be bought and planted at any time of the year, but you will need to water your new plant regularly over the summer if planting now.

Plant the hybrid teas and shrub roses in an area of the garden that receives at least four hours of direct sunlight each day in the summer – ramblers will cope with a little more shade. Also check rose nursery websites for the conditions that each variety prefers, as some have been bred to cope with lower light conditions. Roses also enjoy free-draining soil that does not dry out completely during the summer. Add well-rotted organic farmyard manure or homemade compost to improve the soil where you plan to plant your rose.

Dig a hole twice the width and the same depth as the root ball. You can add mycorrhizal fungi (e.g. Rootgrow) into the hole, which will help the roots to establish, but mine seem to grow well without it. Tease out any congested roots if you have purchased a container-grown plant and place the rose in the hole, making sure that the graft union (the knobbly section at the base of the stem) is at soil level and not buried below the surface.

Back-fill gently with the excavated soil, water well, and add an organic mulch such as shredded bark or homemade compost over the root area, leaving a 10cm (4in) gap free of mulch around the stems.

Aftercare

Water your young rose regularly during the first two summers after planting it, and any prolonged dry spells in spring. You can also add a rose fertilizer each spring, but this may not be necessary if your soil is in good health.

CHAPTER 7

JULY

Summer is in full swing now and my garden is filled with a kaleidoscope of colours, textures, and scents. The bees are everywhere, too, filling me with excitement and joy. I've learned over the years to slow down during the summer months and recognize the benefits of living in the moment.

There is so much in the garden to feed the soul at this time of the year – from picking the fresh herbs I grow by the kitchen door, to creating a simple bouquet using the flowers from my cutting garden and borders. The berries are also starting to yield their jewel-like fruits now, and taste all the better when eaten fresh from the bush.

My beautiful lavender hedge, which I grew from cuttings a few years ago, has been in full bloom for a few weeks now, reminding me of wonderful holidays in Italy, listening to nightingales, walking through meadows, and discovering amazing butterflies. Try planting a flower or shrub that you associate with special times and I guarantee it will lift your spirits every year when it blooms.

Clockwise from top left: the tall spires of blue *Veronicastrum* flowers are loved by bees and butterflies; use flat-topped yarrow blooms to contrast with spiky flowers; the fiery red *Crocosmia* 'Lucifer' glows from the borders; sweet peas grown last autumn from seed continue to produce their fragrant blooms throughout July.

IN SEASON

The garden waxes and wanes in summer, as the early flowers are replaced with a whole new set of blooms. I love the rhythm of the garden as the weeks pass, but it's important to plan ahead if you want a continuous show later in the year.

WHAT'S IN BLOOM?

My July garden is packed with colour from a whole range of perennials and annuals that I have grown from seed, alongside the shrubby hydrangeas and roses (see also p.121 and p.91), which add structure to the borders. My lavender hedge (*Lavandula* x *intermedia*) is a sea of blue, too, and the dahlias are also coming into their own, filling the garden with their exquisite blooms (see p.58).

Montbretia (*Crocosmia*)

I grow the tall, bright red *Crocosmia* 'Lucifer' which shouts from the borders when it appears in July between strappy green foliage. It sometimes needs a little support but there are many shorter varieties to choose from that stand up on their own.

Sweet peas (*Lathyrus odoratus*)

I grow these climbing annuals from seed in the autumn (see p.143) and my reward is the delicate scented flowers that cover my homemade pyramid plant support this month. So much value from an inexpensive packet of seeds – what's not to like?

Gaura (*Oenothera lindheimeri*)

One of my favourite perennials, gaura's tall spikes of small pink or white flowers appear for many weeks in summer. It is tall, reaching up to a metre (3ft) in height, but the slim stems cast little shade over their neighbours, allowing you to plant it throughout the border.

Yarrow (*Achillea*)

The flat-topped flowers of perennial yarrows make ideal partners for plants with contrasting flower spikes or airy grasses. They come in a range of sizes and colours, including white, yellow, terracotta, pink, and red.

Border phlox (*Phlox paniculata*)

Leafy stems topped with rounded heads of flowers in shades of pink, white, and purple will boost the flower count when the border phlox begins to bloom in midsummer. I grow a few varieties, and take cuttings in spring and summer to fill gaps or give away (see p.70).

Brook thistle (*Cirsium rivulare* 'Atropurpureum')

This tall perennial has see-through stems topped with dark pink thistle-like flowers, which sway above my beds like little paintbrushes. Plant them in groups throughout your beds and borders.

Hosta (*Hosta*)

I grow lots of hostas and while some get eaten by slugs, others seem to remain more or less intact, producing their spires of blue, mauve, or white funnel-shaped flowers in July, which I like to include in floral arrangements for the house. They are also very easy to propagate by division in the spring or autumn (see p.146).

Mullein (*Verbascum*)
These tall perennials produce rosettes of leaves, some wonderfully soft, and tall stems of small saucer-shaped flowers. I use them to add vertical accents to my flower border, and they self-seed all over the garden – buy one or two plants and you'll never be without them.

Culver's root (*Veronicastrum virginicum*)
This easy-going perennial is perfect for a late summer border, its tall spires of pale violet-blue or white flowers contrasting beautifully with round or flat-topped plant partners. Bees and butterflies love it too. It will grow in most soils and flowers in full sun or part shade.

MINDFUL MOMENT

When life starts to feel a little overwhelming, perhaps because of the state of the world or challenges at home, I go out into the garden to find some solace. My recipe for staying positive, even in the darkest times of my life, is always the same: go outside, look for pretty flowers, textures, and scents and feed my mind with the surrounding beauty; replace negative thoughts with positive ones; find pleasure in simple things; count my blessings and feel grateful for what I have; create something and photograph nature. I hope some of these ideas will work for you, too.

SPOTTING WILDLIFE

Ladybirds

Loved by children, these cheerful small red beetles are every gardener's friend, helping to control aphids and scale insects, which are their favourite foods. There are more than 40 species in Britain, some with over 20 spots on their backs. Not all are red, either, and you may find yellow, black, brown, or orange species visiting your garden.

In spring, the adult ladybirds lay their eggs on the underside of a leaf, and these then hatch into bullet-shaped larvae that look quite unlike the adults, but they also feast on aphids and other small insects. It's worth looking up an image of these little creatures so you do not mistake them for unwanted insects. After feeding for a few weeks, the larvae pupate, and the adults emerge in late summer. Ladybirds then hibernate in winter in crevices and plant stems in the garden, so it is important not to tidy up too much before the cold weather sets in, and to provide them with suitable hiding places. Also plant pollen-rich flowers – look out for the Plant for Pollinators logo on labels – as these will attract adult ladybirds.

MONEY-SAVING TIP

Working in harmony with nature can save you money on pest control. As well as ladybirds and their larvae, other useful creatures include frogs and toads, birds, hoverfly and lacewing larvae, and earwigs (which eat fruit aphids) that collectively help to control slugs, snails, caterpillars, and aphids that eat our precious plants. To attract this army of helpers, plant a diverse range of pollen-rich flowers that bloom throughout the year to support the insects, and trees, large shrubs and plants such as grasses that produce berries and seedheads to support the birds. A small pond will provide a breeding ground for frogs and other amphibians.

Ladybirds and their larvae are all excellent pest-predators, feasting on aphids and scale insects that threaten our plants. The non-native harlequin ladybird, originally from Japan, is orange-red with black spots, and research is ongoing into the harm it may, or may not be, causing native populations, such as this one.

Clockwise from top left: sow seeds of perennials such as brook thistle (*Cirsium*) to produce plants for next year's borders; mown paths and patches of long grass provide food and habitats for wildlife; wait until meadow flowers have set seed before cutting them; take semi-ripe cuttings from non-flowering stems of phlox.

WHAT TO GROW IN JULY

July is a great month for sowing perennial seed, which will germinate and grow quickly in the warmth and high light levels. I also continue sowing biennials such as foxgloves and sweet Williams to flower next spring, and some bedding for my autumn and winter pots.

SOW INDOORS

Flowers

Sow biennials and perennials (see p.25) under cover in July to protect seedlings from pests.

Herbs

Continue to sow small batches of annuals such as basil and coriander.

SOW OUTSIDE

Flowers

Continue to sow hardy annuals, biennials, and pansies for flowers up to the frosts and beyond.

PROPAGATE NOW

Take semi-ripe cuttings

Use this method (see p.100) to propagate climbers, herbs, perennials, shrubs, and trees.

Root in water

Many plants will root in water (see p.116).

PLANT NOW

Cuttings taken earlier

Continue to plant out cuttings taken earlier, keeping them well watered until established.

Perennials

You can continue to plant out perennials grown from seed or new plants but try to avoid doing this just before a holiday, as these young plants will need watering every day or two in hot weather.

Herbs

Continue to plant out annual herbs grown from seed or cuttings, including chives, basil and coriander and dill.

MAINTENANCE MUSTS

Water bedding and young plants

Young plants are vulnerable to drought in summer and need watering daily, so if you do not have kind neighbours to help when you go on holiday, consider installing an automatic watering system to keep them alive.

Mow the meadow

We mow our meadow at the end of July, after most of the flowers have set seed, then leave the mowings on the ground for a few days to allow any further seeds to fall on to the soil. We then collect up all the mowings, and use the straw as mulch or to make compost.

Collect seeds

Many early-flowering plants will be scattering their seeds in July for you to collect and sow to make more plants for free (see p.103).

ANYA'S PROPAGATION MASTERCLASS

SEMI-RIPE CUTTINGS

Remove stems with a hard base and soft tip for semi-ripe cuttings, which can be used to propagate evergreens, shrubs, mature perennials and a whole variety of other plants.

July is a great time to take semi-ripe cuttings from the plants in your garden. Similar to softwood cuttings, these are taken from more mature stems and are a little more robust. I particularly like to propagate the little conifers from my hedge to use in my winter pot displays. These tiny trees soon root, ready to be planted out the year after, saving me lots of money. Also try this method for the wide range of other plants listed below.

You will need

Clean, sharp knife or secateurs
Recycled pots
Peat-free cuttings compost or peat-free multipurpose with horticultural grit.

1. Choose non-flowering stems that have grown this year, removing them at the base. Keep them in the refrigerator if you can't prepare the cuttings immediately.

2. Trim each stem to 10–15cm (4–6in) in length, cutting just below a leaf. Then remove the lower leaves, so that you have about four leaves or sets of leaves on the stem. When taking cuttings from plants with large foliage, such as hydrangeas, cut each leaf in half.

3. Insert the cutting into a pot of free-draining compost such as a 4:1 mix of peat-free multipurpose and horticultural grit. Water well, and leave to drain. I keep my cuttings in a sheltered spot, out of direct sun, in the garden but you can pop them on a windowsill out of direct sun if you prefer. Keep the compost damp but not wet.

4. When new top growth appears, look through the holes at the bottom of the pot for new roots. Then tip out the cuttings and transfer them to pots of their own.

Heel cuttings

These cuttings are prepared in the same way as semi-ripe cuttings, but the side-stems are removed with a little tail or 'heel' of bark from the main stem still attached. Growth hormones beneath the bark in the heel promote rooting, so try this if you don't have success with the basic method.

When to take cuttings

Take semi-ripe cuttings in summer and autumn, from July to September.

Suitable plants

- Shrubby herbs, such as sage, curry plant, lavender, and rosemary
- Mature perennials
- Shrubs
- Conifers

1. 2. 3. 4.

1.

2.

3.

4.

ANYA'S PROPAGATION MASTERCLASS

SOWING HARVESTED SEED IN POTS

Packets of seeds are not expensive, but those harvested from your own flower pods are completely free. Check all your flowers and collect these tiny treasures from ripe seedheads.

I love wandering around my garden in July with a few paper bags, collecting seed from my favourite plants, including gaura (*Oenothera lindheimeri*), shown here. The pods are usually ready to harvest about two months after flowering when they turn brown or beige.

You can also take seed from berried plants, too, when they are ripe but before the birds get to them. Squash the berries in a fine sieve to expose the seeds and then run them under the tap to remove the flesh, before leaving them on paper towels to dry.

You will need

Clean, sharp secateurs
Paper bags or airtight containers
Marker pen
Plate or tray
Seed trays or pots
Peat-free seed compost

1. Look for ripe seedheads, which have turned beige or brown. Cut off the heads and pop them into paper bags or an airtight container. Use a marker to label them. Then gently prise open the pods over a plate or tray and tip the seed on to it. Remove any bits of the seed pod and other material, known as 'chaff', which could harbour disease.

2. Leave the seeds to dry for a day or two. Then follow the instructions on p.25 for sowing seed into pots of seed compost. Once the seedlings have a few leaves, carefully transplant them into individual pots or module trays (as shown).

3. Keep the pots or module trays in a sheltered spot outside (if seedlings are hardy) or in a greenhouse or on a windowsill inside.

4. Keep the seedlings moist but not wet as the roots establish. The plants should be well established the following year, when you can plant them outside.

When to collect seed

Seedheads will be ripening throughout the growing season, from spring to late autumn.

Suitable plants

Many annuals, perennials, and herbs produce viable seed, but some of my favourites include:
Marigolds (*Calendula*)
Foxgloves (*Digitalis*)
Sweet peas (*Lathyrus odoratus*)
Love-in-a-mist (*Nigella*)
Gaura (*Oenothera lindheimeri*)
Hellebore (*Helleborus*)

PROJECT OF THE MONTH
ALPINE PLANTER

This pretty planter is very easy to make and because the plants used are alpines, they are adapted to drought in summer and need very little aftercare. They are also hardy and can be kept outside all year in a sheltered spot.

This is a very inexpensive project since many alpines, such as houseleeks and sedums, grow quickly and multiply easily from offsets (see p.54). I have created this beautiful display from my existing plants and a planter given to me by my mother-in-law, but if you don't have any plants, buy a few to plant now and propagate them to make more.

You will need

Planter with drainage holes
Peat-free multipurpose compost
Horticultural grit
Selection of hardy alpines; I have used houseleeks (*Sempervivum*) and a small sedum (*Hylotelephium*)

1. Use a pot with drainage holes in the base or drill some if yours doesn't have any. Then fill it with a 3:1 mix of peat-free compost and horticultural grit. Carefully remove a plant from its pot – I'm using those I propagated from offsets in a tray.

2. Plant the alpines, making sure none are touching and the leaves are not buried under the surface, which may cause them to rot.

3. Fill a small jug with some more grit and pour it onto the compost surface between the plants. This helps to prevent the soil splashing up onto the leaves when it rains and adds a decorative finish.

4. Set your alpine planter where you can admire the intricate details of the plants – I put mine on a garden table or beside a garden chair, so I can admire it from above.

Aftercare

These plants are very drought resistant and only need watering once a fortnight, even in summer. You can leave the planter outside over winter, too, since they are all hardy.

Suitable plants

Any low-growing sedum species and houseleeks are suitable, including the cobweb houseleek, *Sempervivum arachnoideum*, which is covered with hairs and a great talking point.

EASY MONEY-SAVING TIP

Plant these shallow-rooted alpines in a repurposed baking tin, large seashell, tool box or other item that you can drill holes into the base, instead of buying new containers.

Clockwise from top left: French lavender (*Lavandula stoechas*) is the first to flower in late spring; my lavender hedge is a magnet for bees in summer; grow the compact *Lavandula angustifolia* 'Munstead' as a low hedge; *Lavandula* x *intermedia* 'Edelweiss' and *Lavandula angustifolia* 'Hidcote' combine beautifully in a border.

FOCUS ON LAVENDER

This versatile plant has so many uses in the garden, its colourful flower spikes and aromatic foliage making beautiful hedges, border specimens, and summer container displays.

There are many different species of lavender to choose from and I love them all. Some are short and perfect for pots, others are taller with long flower stems that need space to thrive, so check heights and spreads before buying to ensure they suit your needs.

HOW TO GROW LAVENDER

The best time to plant lavender is spring, when the warmer temperatures will help it to establish before winter. It needs a sunny site and free-draining soil – if yours is heavy clay, grow it in a large container of peat-free loam-based compost mixed with horticultural grit.

When planting a hedge, space smaller varieties such as *Lavandula angustifolia* 'Munstead' about 30cm (12in) apart, and large types, including *Lavandula* × *intermedia* 'Grosso', about 45cm (18in) apart.

Aftercare

Although lavender is drought resistant once established, it will need watering regularly after planting. Water those in pots every week in spring and summer.

Lavender can become woody at the base and produce fewer flowers as the years pass. To extend its life, prune it by mid-September or in spring, cutting each plant into a compact, round shape, while leaving plenty of new shoots. Replace any plants that do become woody and unproductive.

Plants for free

Lavender is very easy to propagate and I take softwood cuttings (see p.70) every year to replace old plants that are past their best. French and English lavender (*L. stoechas* and *L. angustifolia*) is also quite easy to grow from seed, which you can collect from your existing plants in late summer. Sow the seed straight away or in spring, in seed compost or a mix of 40 per cent peat-free compost, 30 per cent sharp sand, and 30 per cent horticultural grit. Cover the seeds with a sprinkling of soil or horticultural grit and spray with water, rather than watering from a can, ensuring the compost is damp but not wet. Keep in a warm room or greenhouse and the seeds should germinate within 5 to 30 days.

FAVOURITE VARIETIES

There any many wonderful varieties available, but my favourites include *Lavandula angustifolia* 'Hidcote', a compact variety, ideal for low hedges; *Lavandula* × *intermedia* 'Phenomenal', a tough medium-sized lavender, ideal for hedging, and *Lavandula* × *intermedia* 'Grosso', which, as the name suggest, is a tall form with spikes of highly scented blue-purple flowers in summer. If you want a white form, *Lavandula* x *intermedia* 'Edelweiss' is a good compact variety, ideal for pots.

CHAPTER 8
AUGUST

The garden in August is filled with colours, textures, and insects, and the summer holidays offer a break from work and school to get creative. When I was growing up, I remember my sister and I were always making things – arranging flowers, drying stems for long-lasting bouquets, creating seasonal gifts – all from our garden.

I still enjoy looking for natural treasures and gathering souvenirs from day trips and holidays that remind me of these special times with my family. A rock, shell, or a picture of a beautiful piece of driftwood on the beach, are far more meaningful than items you can buy in the shops. Going on a nature hunt will inspire you to look for more peaceful moments, too, while making something beautiful with your found objects will also bring a great sense of achievement.

There are also lots of seeds to collect now and many, such as honesty (*Lunaria annua*) and foxgloves (*Digitalis*) can be simply scattered around the garden, saving you money with the minimum of effort.

Clockwise from top left: white *Dianthus* MEMORIES and 'Gran's Favourite' both produce a sweet clove scent throughout summer; the coppery heads of *Helenium* 'Moerheim Beauty' light up the garden; the beautiful white admiral occasionally visits our garden: I grow *Achillea ptarmica* (The Pearl Group) 'The Pearl' from seed.

IN SEASON

Summer is now coming to end and the flurry of growth seen earlier in the season is starting to slow as the seasons turn. I like to plan ahead in late spring and include plants that will provide plenty of colour in my beds and borders now.

WHAT'S IN BLOOM?

Hydrangeas continue to adorn the garden with their big, blowsy blooms (see p.121), while asters and heleniums are starting to unfurl, too. Keep deadheading your roses, dahlias, and summer bedding to maintain the show throughout August and beyond.

Salvias

This versatile group of summer perennials look their best in July and August, decorating the garden with their spikes of small flowers. My favourites include the inky blue *Salvia* 'Amistad'; violet *Salvia nemorosa* 'Caradonna'; and pinky purple *Salvia* 'Love and Wishes'.

Pinks (*Dianthus*)

These sweetly scented little beauties are perfect for the front of a border, and I've always grown a few in my garden to remind me of my childhood home. Remember to deadhead them regularly to continue the show into August and take cuttings from non-flowering shoots to increase your stocks.

Verbena bonariensis

An all-time favourite, this tall perennial is so easy to grow from seed and will self-seed around the garden, too, in free-draining soil or a gravel bed. Its small purple flowers have a sweet clover scent and attract pollinators.

Agapanthus

I wouldn't be without my agapanthus in late summer. The large domes of small blue or white trumpet-shaped flowers, which appear on sturdy stems over strappy foliage, are perfect for patio containers or a sunny border, and they flower for many weeks.

Helenium

Flowering from August until mid-autumn, the russet tones of heleniums are like little lamps in the border, brightening up dull areas with their cone-shaped blooms. Plant them in full sun for the best display.

Achillea ptarmica (The Pearl Group) 'The Pearl'

Most achilleas have flat-topped flowerheads but I grow this beauty for its dainty button-shaped white flowers, which appear on airy stems throughout the summer. It is a good choice for the cutting garden and easy to grow from seed (see p.25).

Morning glory (*Ipomoea tricolor*)

A gorgeous annual climber, morning glory needs a sunny site to flower – the blooms open in the morning, then die in the afternoon, but then produce new flowers the next day, hence the name. There is a wide colour choice, and it's easy to grow from seed in early spring.

Catmint (*Nepeta*)
This perennial plant produces a succession of lavender-blue flowers, which bees and pollinators adore, from early to late summer, making it exceptional value for money. The low-growing types such as *Nepeta racemosa* 'Walker's Low' are perfect for edging a border, and all forms can be propagated from cuttings.

Great burnet (*Sanguisorba officinalis*)
I love this plant's little dark red or pink flowers that dance in the breeze on slender stems over ferny foliage. The flowers bloom for many weeks in summer and early autumn, and you can divide clumps (see p.146) or taking cuttings to make more plants for free.

MINDFUL MOMENT

Every winter, I start my hunt through seed catalogues and online nurseries for some new plants to sow, and take great joy in seeing them germinate, put on growth, and finally set the garden alight with their joyous flowers. The annuals always put on an amazing show and are often at their best in August, when I can take a little time off and simply sit with my family or have a quiet moment on my own with a coffee, surrounded by the plants I've grown. It's so satisfying, and you can do this too. Make a note in your diary or on your phone to order some catalogues in December or January and use those dark days to find inspiration and imagine the promise those tiny capsules hold for the months ahead.

SPOTTING WILDLIFE

White admiral butterfly
We are very lucky to see lots of butterflies in our garden during the summer months, but when the white admiral flies in for a visit, we feel exceptionally blessed. This medium-sized butterfly has dramatic white marking on its inky black wings and it often glides rather than flutters. It prefers shady woodlands, which we have close by, and you may spot it feeding on bramble flowers in the spring.

The adults lay their eggs on honeysuckle leaves, which provide food for the bright green caterpillars with brown spines that emerge soon after. In August or early September, the caterpillars make a silk shelter on a leaf where they pupate to sit out the winter, the adults finally emerging the following June as the temperatures warm up. To encourage these elegant insects into your garden, plant some honeysuckle close to trees, if you have any.

ANYA'S MONEY-SAVING TIP

I take softwood or semi-ripe cuttings (see p.70 and p.100) of tender perennials such as *Salvia* 'Amistad' and pelargoniums in August, growing them on and overwintering them, so that they are ready for next year's pots and garden borders. I also take cuttings of short-lived perennials such as pinks (*Dianthus*) now. Thinking ahead saves money because all of these plants are expensive to buy in pots at the garden centre and this self-sufficient approach also makes me feel like I'm positively contributing to the environment by using what is available in my garden.

Right Propagate herbs throughout the summer by rooting them in water (see p.116) or taking semi-ripe cuttings (p.100). I've raised a whole host of them in this way for my garden.

Clockwise from top left: keep watering and feeding Mexican fleabane (*Erigeron karvinskianus*) and other pots of flowers; take semi-ripe cuttings from *Verbena bonariensis* and other perennials now; sow *Phacelia* to improve your soil; biennials such as sweet Williams can be sown now for flowers next spring.

WHAT TO GROW IN AUGUST

August is a quiet month for sowing and planting, but you have to keep on top of watering to ensure any young growth does not wilt during the warm weather. Also continue to feed tomatoes, fruits and flowers in pots and containers.

SOW INDOORS

Flowers

It's not too late to sow winter bedding plants such as violas and pansies at the beginning of August, but delay sowing if you are going on holiday this month. You can also sow hardy annuals such as larkspur (*Consolida*), cornflowers (*Centaurea cyanus*), and California poppies (*Eschscholzia californica*), as well as biennials that will flower next year.

Herbs

Continue to sow annual herbs such as basil and coriander for autumn harvests.

PROPAGATE NOW

Semi-ripe cuttings

Continue to take these cuttings throughout August (see p.100).

Rooting in water

Add prunings from shrubs that bloomed earlier in the year in water to root (see p.116).

PLANT NOW

Perennials

Continue to plant perennials grown from seed or cuttings if you can water them frequently. If you are away, delay planting until you return.

Herbs

Continue to plant out annual herbs grown from seed or cuttings.

MAINTENANCE MUSTS

Summer prune wisteria

Prune back the long whippy stems that grew this year to five or six leaves to keep the plant size in check and to encourage more flowers to form next year.

Keep ponds and water features topped up

If you have a wildlife pond, use rainwater from a butt to top it up, or fill a few buckets with tap water and leave it for a day to allow the chlorine to evaporate.

Feed the soil with green manures

Sow *Phacelia* on bare soil to smother weeds and offer pollinators late-season flower feasts. Just before it sets seed in mid-autumn, dig the plants into the ground, which enriches the soil with the plant nutrient nitrogen, increasing its fertility for next year's flowers and crops.

Trim hedges

Cut your hedges at the end of August, but only when you are sure that any baby birds have flown their nests.

ANYA'S PROPAGATION MASTERCLASS

ROOTING IN WATER

This month, I want to show you a very easy propagation method to do today. Rooting the stem tips from a whole range of annuals, perennials and shrubs is almost foolproof, requiring no special equipment, so experiment with whatever you want to grow.

This is the easiest method of propagation and all you will need to do it is a few jam jars or vases and a windowsill.

I use this method when taking cuttings from my own garden and also when I visit friends who may have a plant I would like to grow. Many herbs, annuals, perennials, and even shrubs will root easily in a jar of water. I also use this method to increase fruit crops such as redcurrants, and, later in the season, I keep a supply of tender pelargoniums for the following year's displays by rooting some stems in water in early autumn, then potting them up and overwintering them in a frost-free place under cover until late spring.

You will need

Sharp, clean secateurs or scissors
Jam jar or vase
Recycled pots
Peat-free cuttings compost or peat-free multipurpose compost and grit

1. Remove a non-flowering shoot from your plant, just below a leaf or side-stem. It doesn't really matter how long the shoot is, but aim for between 10 – 20cm (4–8in).

2. Cut off the lower sets of leaves so that none will be submerged in the water. Then plunge the stems in a jam jar or vase filled with fresh tap water. You can cover the vessel with chicken wire before adding your cuttings to hold the leaves above the surface.

3. Refresh the water every few days and when roots appear, pot on your stems into containers of peat-free cuttings compost or multipurpose mixed with a handful of grit.

4. Grow on in good light in a frost-free place, if taking cuttings from tender plants, or a sheltered spot outside if your plants are hardy. Keep the compost moist but not wet.

When to take cuttings

Any time of year.

Suitable plants

Over the years, I have tried rooting a whole range of annuals, perennials, herbs, and shrubs in this way, and most have been successful, so I would recommend experimenting with whatever you have to hand.

Clockwise from top left: most herbs (top left) will root easily in water, and I have also propagated shrub prunings (top right) in this way; placing chicken wire over the jam jar holds the leaves above the water surface; try taking cuttings from cosmos and other annuals in spring to increase your supply of summer container plants.

PROJECT OF THE MONTH

GROW CHILLIES YEAR-ROUND

I always have a few pots of chillies that I've grown from cuttings on my kitchen windowsill, providing me with pretty flowers followed by spicy fruits from midsummer to winter.

Chilli plants can be expensive to buy in the spring but they are very easy to propagate, so once you have a plant, you will never need to buy another, unless you want a different variety, of course.

Most experts advise growing chillies from seed as annuals as a cheaper alternative to buying plants. However, I find that these short-lived perennials are very easy to grow all year round from cuttings. My propagated plants also mature more quickly than seed-raised chillies and allow me to harvest the fruits earlier and for longer. Just remember that chillies are tender, so grow them indoors throughout the year.

You will need

Sharp, clean knife or secateurs
Plant pot
Cuttings compost or peat-free compost mixed with some horticultural grit

1. Select a few non-flowering stems about 15–20cm (6–8in) long, and cut them just below a leaf node (see p.86). Remove the lower few sets of leaves.

2. Place your cuttings in a pot filled with compost and water them well.

3. Store your pot of cuttings indoors, out of direct sun, until you see new shoots develop, usually between 3–6 weeks later. Then pot up each of the young plants into a container of their own.

4. After repotting, pinch out the tip of the main central stem of each plant. This will promote more sideshoots to develop and the bushier plants will then produce more fruits.

Aftercare

Keep the compost moist, but not wet, and feed plants with a high potash fertilizer such as tomato feed every fortnight when the flowers appear. Stop feeding in late autumn when plants become dormant.

TOP GROWING TIP

Chillies need heat and light to produce a good crop, and growing them in a greenhouse or conservatory is best. They can be also grown on a sunny windowsill – turn your plants regularly to prevent them growing towards the light. Overwinter chillies on a windowsill, too, if you do not have a heated greenhouse.

Clockwise from top left: *Hydrangea arborescens* 'Annabelle', with its large white flowerheads, is one of my favourites; the pale pink hydrangea (top right) and darker pink one (bottom right) were both raised from cuttings and are now big, established shrubs; *Hydrangea paniculata* produces large cone-shaped flowers in late summer.

FOCUS ON HYDRANGEAS

Hydrangeas are like garden goddesses, their large heads of white, pink, purple, or blue flowers putting on a magnificent show from midsummer to the first frosts, after which their decorative seedheads can be cut for indoor displays.

There are a few different types of hydrangea to choose from, and while some grow into very large shrubs, others are more compact and bred for growing in a container, so check plant labels carefully for plants to suit your space. Blue hydrangeas will only retain their colour when grown on acid soil, or in ericaceous compost, so use a kit from the garden centre to test your soil type – the flowers will turn pink if yours is alkaline.

How to plant hydrangeas

The best time to plant hydrangeas is in early spring or autumn. Choose a site with moisture-retentive soil – they dislike dry roots – in full sun or part shade. Improve the soil by digging in well-rotted homemade compost or manure from an organic source. Dig a hole three times as wide and the same depth as the root ball, and ensure the stems are at the same level as they were in their original pot after planting. Water well, and add a mulch of chipped bark or homemade compost over the root area, leaving a gap around the stems, which will help to retain moisture in the soil.

Aftercare

Keep plants watered until they are established. Continue to water those in pots during dry spells and feed with a homemade fertilizer (see p.52) from spring to early autumn.

Pruning hydrangeas

All hydrangeas benefit from annual pruning in spring, but find out which type you have before you start cutting. *Hydrangea arborescens* and *Hydrangea paniculata* both bloom on new wood and are pruned in March. Either cut all the stems down to a pair of healthy buds just above the ground to produce plants with fewer stems and very large flowers, or remove one third of the length of each stem, again cutting to a pair of healthy buds, for more stems with smaller flowers. I prefer the latter method, as the flower stems are less likely to need support.

Prune mopheads and lacecaps less severely in late March, cutting them back to the first or second set of healthy buds from the top of each stem. If you prune any harder, you will remove the flower buds, which will have formed the previous year.

HYDRANGEAS FOR FREE

When pruning hydrangeas, pop non-flowering cut stems in a vase of clean water. New shoots will soon appear, and you can then take heel cuttings from these (see p.100). Remember to cut the foliage in half before inserting the cuttings into pots of peat-free compost mixed with horticultural grit. Grow on, keeping the compost moist, but not wet.

CHAPTER 9

SEPTEMBER

September marks the beginning of autumn and a busy time in the garden. There are lots of seeds you can sow now and many propagation jobs to be getting on with. Sharing these tasks with my growing family makes them even more enjoyable, while relishing the last of the summer's warmth as we work.

I've been gardening with my boys since they were very little and witnessing the joy they get from seeing a tiny seed develop into a mature plant. Growing with children is like planting positivity in their hearts, helping them to understand sustainability and how to manage resources such as water and energy.

While they may lose interest in gardening as they get older, the seeds planted in their minds about the beauty and benefits of Nature may germinate again later in their lives and, as adults, they will remember the times they spent outside with their parents and grandparents. They may even want to share this knowledge with their own children. These are the simple pleasures we have on our doorstep that everyone can enjoy.

Clockwise from top left: Japanese anemones flowering in early September; Geranium ROZANNE blooms from early summer right through to the autumn when deadheaded regularly; *Rudbeckia subtomentosa* 'Henry Eilers' is an eye-catcher; the blue-flowered aster, *Symphyotrichum* 'Little Carlow' is one of the best.

IN SEASON

The garden is still full of flowers in September, with earlier blooms replaced by tall perennial daisies such as cone flowers and asters, which deliver a sea of beautiful colours while also supporting bees, butterflies, and moths.

WHAT'S IN BLOOM?

Plan your garden to produce a succession of blooms right through the autumn. I plant these beauties in my borders to create a full spectrum of colour at this time of year. Repeat-flowering roses and dahlias also continue to sparkle in September, if you dead-head them regularly.

Japanese anemone (*Anemone hupehensis* and *Anemone japonica*)

These tall perennials put on their show of white or pink round flowers from midsummer to early autumn, the blooms appearing on slender stems above lobed green foliage. *Anemone hupehensis* is easy to grow but it's very vigorous and may take over a border, so opt for the better-behaved *Anemone japonica* and its cultivars in smaller spaces.

Coneflower (*Rudbeckia*)

This group of plants produce eye-catching yellow daisy flowers in September and come in all sizes, from compact types for the front of the border to the tall, waving stems of *Rudbeckia laciniata* 'Herbstsonne', which reach almost 2m (6ft 6in) in height. Another favourite of mine is the appropriately named sweet coneflower, *Rudbeckia subtomentosa* 'Henry Eilers', which looks like a child's drawing of the sun, with slim yellow petals radiating out from a brown central cone.

Monkshood (*Aconitum carmichaelii*)

The elegant spikes of dark blue hooded flowers of this *Aconitum* decorate the garden up to the end of the month. It's useful because it will bloom in fairly deep shade, but like all monkshoods, all parts of the plant are poisonous, so it's not a good choice for gardens used by young children or pets.

Asters (*Symphyotrichum*)

My favourite asters include *Symphyotrichum* 'Little Carlow', which produces masses of violet-blue daisies on tall stems throughout autumn, and the New England aster *Symphyotrichum novae-angliae* 'Andenken an Alma Pötschke', with its bright cerise blooms. The small-flowered pink *Symphyotrichum* 'Coombe Fishacre' is also on my wish-list and blooms for many weeks up to mid-September.

Honeywort (*Cerinthe major* 'Purpurascens')

An unusual looking hardy annual with glaucous leaves and drooping purple-blue flowers, honeywort will self-seed freely around the garden. I collect and sow seed at intervals throughout the summer, which prolongs the flowering display into early autumn.

Hardy geraniums (*Geranium*)

I have a range of these stalwarts, and together they flower from May to September. In September ROZANNE continues to pack a punch, this amazing little plant offering incredible value for money. Divide plants in spring or autumn to increase them (see p.146).

Feather reed grass (*Calamagrostis × acutiflora* 'Karl Foerster')
I love this tall ornamental grass in autumn when the stems of seedheads rustle in the breeze and add texture to the flower garden.

Nerines
The spidery blooms of these autumn-flowering bulbs are hard to resist. Choose the easy-care hardy *Nerine bowdenii* types, with their white or pink flowers, and plant the bulbs in summer with the tips just above the surface. To perform well, choose a site with well-drained soil at the base of a sunny wall for added protection, since these South African natives are vulnerable to hard frosts in cold areas.

MINDFUL MOMENT

September is when the light starts to change, softening from the harsh rays of summer and enriching the colours of the flowers. Over the years I have learned to embrace all seasons and not to feel sad about the colder months to come. In fact, I love this time of the year for many reasons. The garden looks so beautiful and peaceful, while the variety of textures created by grasses, shrubs, and flowers feed my mind and soul. The pollinators are still busy, too, darting around in the early autumn sun, feeding on my lavender and daisies. I like to take a moment to enjoy the dahlias that are now still blooming, the stems of gaura dancing in the breeze and my *Molinia* catching the magical, slanting light in the evenings. Every season has something precious to offer.

SPOTTING WILDLIFE

Pipistrelle bat
Living close to woodland, we are lucky to have bats visiting the garden in the summer and early autumn. Among those that come are the little pipistrelles, the most common of Britain's 18 species of bat. These small nocturnal mammals make their homes in a wide range of habitats, including mature woodlands, wetlands, grasslands, farms, parks, and gardens. They particularly like open grassy areas surrounded by trees or bushes, which is exactly what we have here in our garden.

During the summer months, pipistrelle bats gather in large colonies, roosting under the bark or in hollows in trees, or in the roof spaces of buildings. They are predators and eat a wide variety of flying insects, including flies, midges, and mosquitoes, helping to keep these unwanted creatures in check.

Pipistrelle bats hibernate from mid-October, when they gradually stop feeding and look for somewhere safe to roost over winter, although they will come out on mild days to feed. Pipistrelles may be widespread but their populations are in decline due to the overuse of pesticides, which kills their insect prey, and the loss of their habitats. It is illegal to harm bats, since all species are protected by law in the UK, and don't worry if they make a home in your roof or outbuildings as they will do no damage to them. To support them, include a range of pollen-rich plants to attract their prey and avoid using pesticides.

Right My hydrangeas continue to flower well into the autumn, providing me with beautiful cut stems for my indoor arrangements.

Clockwise from top left: take semi-ripe cuttings from hydrangeas before the cold weather arrives; sow annual poppies in the ground for flowers next year; plant out biennials such as foxgloves that you sowed earlier, for flowers next May or June; rooted *Verbena* cuttings can be planted outside now.

WHAT TO GROW IN SEPTEMBER

Early autumn provides the perfect conditions for root growth and seeds of plants that will flower next year germinate quickly when sown now. It's also time to take cuttings of tender plants, so that they establish in their pots indoors before the cold weather arrives.

SOW INDOORS

Flowers

As temperatures cool but light levels remain relatively high, many perennials will germinate quickly from seed in early September. The cost of seed is low compared to young plants grown in pots, and will save you money, even if just a few germinate. Try these now:

- Globe thistle (*Echinops*)
- Sea holly (*Eryngium*)
- Euphorbias
- Red hot pokers (*Kniphofia uvaria*)
- Lupins (*Lupinus*)
- Verbascums

Herbs

Continue to sow annuals such as basil and coriander for a late autumn harvest.

SOW OUTSIDE

Flowers

Hardy annuals such as those listed here will produce robust plants that will overwinter and flower in late spring or early summer:

- Bishop's weed (*Ammi majus*)
- Pot marigold (*Calendula*)
- Love-in-a-mist (*Nigella*)
- Poached-egg plant (*Limnanthes*)
- Opium poppy (*Papaver somniferum*)

PROPAGATE NOW

Softwood and semi-ripe cuttings

Both can be taken now but they will need to be kept warm indoors over winter. Or simply add the stems to water to root if you don't have time for cuttings (see p.116). Pay attention to plants that will not overwinter outside or are on the tender side, including:

- Salvias
- Pelargoniums
- Chrysanthemums

PLANT NOW

Spring-flowering biennials

Plant out violas, foxgloves (*Digitalis*), and wallflowers (*Erysimum*), sown earlier in the summer, in pots for the patio or into your borders. Think about combining them with bulbs to increase their impact or extend the colour show next spring and early summer.

Spring bulbs

Plant bulbs, which are now available at garden centres and online, for a colourful display next spring. Try daffodils (*Narcissus*), alliums, grape hyacinths (*Muscari*), crocuses, and tulips, and bulbs you've never grown before. Some may be on sale later in autumn, but buy those you love now in case they sell out.

Perennials and climbers

September is a great time to plant new hardy perennials and climbers, or plants grown from cuttings earlier in the year, since the soil is warm and the roots will establish quickly.

Herbs

Plant hardy shrubby herbs such as rosemary, sage, oregano, and thyme in beds and outdoor containers in early September, so that they will establish before the cold weather arrives. Make sure you choose a sunny, sheltered spot with free-draining soil – plant in containers of peat-free compost if you have heavy clay.

MAINTENANCE MUSTS

Start a compost heap

Any month is a good time to start a compost heap, but in September, when you are deadheading and cutting down spent stems, there will be plenty of material to fill one. Layer the contents with green, leafy materials rich in nitrogen between straw-like brown twigs and dried stems, which add carbon to the mix. If you have too much green stuff and not enough brown, add shredded cardboard to your heap to prevent it from getting too wet.

Chop and drop

When pruning back old flower and shrub stems, you can simply cut them into short lengths and spread them in a layer around the plant, rather than composting them. These prunings make an effective mulch that will help the soil to retain moisture and keep weeds at bay. They will eventually rot down and help to condition and feed the soil, too.

Take seeds from perennials

Many early-flowering summer perennials will be setting seed now, which you can either leave to spread themselves around the garden and see what pops up, or harvest them to sow more accurately into pots or borders. See p.25 for my tips on how to do this.

Plant up autumn containers

I use warm days in September to prepare pots for the colder months to come. I add a layer of compost to the base, then plant some bulbs on top – either one type or two that flower at different times – then more compost to fill up the container to about 5cm (2in) below the rim. Finally, I add some evergreens to keep the show going over winter. My favourites include heucheras and sedges, with bellis daisies and violas that I grow from seed in early summer.

Lift and divide congested perennials

Now that summer is over and many perennials are starting to die down, look around the garden for those that have formed large clumps, which could be divided (see p.146) to make new plants. It's best to wait until after the plants have flowered before lifting them, so delay timing for plants such as rudbeckias and sedums (*Hylotelephium spectabile*).

Keep watering and feeding

Young plants still need watering in September, since they will continue to grow until later in autumn and temperatures can remain high this month. Feed until the end of the month, but then stop until spring arrives.

Clockwise from top left: start making a compost heap with your autumn garden waste; plant allium bulbs in free-draining soil; harvest seed from summer-flowering plants to sow now in pots or scatter around the garden; plant clematis grown from cuttings when they have a well-established root ball.

ANYA'S PROPAGATION MASTERCLASS

AIR LAYERING SHRUBS

While you don't often hear about this propagation method, it is surprisingly easy to do, and suitable for many trees and shrubs. The cuttings produce large new plants that would be expensive to buy.

This is a good propagation method for shrubs and climbing plants that do not root easily from cuttings, but you can try it on most woody stemmed plants.

You will need

Clean, sharp knife or secateurs
Used compost bag and scissors
Garden twine
Peat-free cuttings compost
Recycled pots

1. Choose a healthy stem that's about the width of a pencil or slightly wider. Trim off side-shoots and leaves from a 30cm (12in) section, then use a pair of sharp scissors to scrape off a little bark through a leaf bud to create a little tag, as shown.

2. Carefully wrap the stem section in black plastic – I use an old compost bag – to make a little bag around the stem. Tie it securely at the bottom with some garden twine.

3. Pack moist, peat-free cuttings compost under the bark tag, then pack more compost into the bag to cover the stem, before sealing it at the top with twine. Leave the wrapping in place for up to a year, opening it now and then to check for signs of rooting.

4. When strong new roots have developed, remove the plastic, cut through the stem just below the rooted section, and pot it up into a container of peat-free compost to grow on.

Aftercare

Water, label, and grow on the plant until it is large enough to plant in the garden. I often use the young plants in pots on the patio.

When to air layer

Autumn or spring are the best times to air layer your plants.

Suitable plants

I use this method to increase my smoke bush (*Cotinus coggygria*) and camellias. Others you could try include:

- Chaenomeles
- Daphne
- Forsythia
- Witch hazel (*Hamamelis*)
- Jasmine (*Jasminum*)
- Magnolia
- Rhododendron and azalea
- Lilac (*Syringa*)
- Viburnum

1.

2.

3.

4.

PROJECT OF THE MONTH

CREATING A CUTTING GARDEN

I have made my cutting garden almost entirely from plants either sown from seed or propagated from dahlia tubers and other plants. This money-saving idea will fill both your home and garden with beautiful flowers.

Many cut flowers come from abroad, either flown in from far-away lands or shipped from the Netherlands and Europe, so growing your own at home in the garden helps to reduce your carbon footprint, as well as providing beautiful blooms for your home.

Growing your own flowers is relatively easy, but you will need a sunny site and free-draining soil that retains some moisture for the best results. You can improve the soil's fertility by adding homemade compost or well-rotted manure from an organic source, which will help to improve its structure, but you can't change the amount of sun your garden receives, so choose your spot wisely.

Colours and planting styles

Think of your favourite colours and textures and choose flowers, herbs, and grasses that will create a lovely combination in a bouquet. There are no rules – it's all about growing blooms that will please you.

You can also choose different styles of planting, either growing your flowers in rows like a vegetable plot, or matching the style of your borders and the rest of the garden.

Spacing your plants

When you start planting and sowing, consider the final spread of each of the plants. This information is easy to find on seed packets or online. Then plant them so that they have space to grow to their full potential, with the tallest at the back, and you can access them easily to cut them. I grow mine in rows, leaving about 45cm (18in) between my rows.

Staking and weeding

Cutting gardens are not low maintenance so bear this in mind if you don't have much time. Many cut plants, such as dahlias and delphiniums, will need staking, and the gaps between the plants need weeding regularly.

Recommended plants

When choosing plants to grow for cut flowers, think of those that will offer you a long season of picking. There are hundreds, if not thousands, to choose from, but here are my favourites:

- *Celosia argentea* (Spicata Group) 'Flamingo Feather' – pinkish purple feathery blooms that add texture as well as colour
- Dahlias – grow a good variety
- Hydrangeas – for fresh and dried flowers
- Statice (*Limonium*) – perfect for dried flowers, it comes in pink, purple, and white
- Peonies – great for late spring arrangements
- Roses – those with long, straight stems
- Strawflowers (*Xerochrysum bracteatum*) – bright orange, yellow, and pink papery blooms, ideal for dried arrangements

CHAPTER 10

OCTOBER

You may be surprised to learn that October is my favourite month. While many people may think it's all over in the garden for another year, nothing could be further from the truth. October is full of natural treasures – ripening apples to savour, colourful berries on the trees and shrubs, leaves putting on their fiery autumn cloaks, and the last blooms shining out from the borders. And because the days are now shorter and the weather can often keep me indoors, my time in the garden is even more precious.

When I was a little girl, someone gave me a kaleidoscope. I found it fascinating. I would sit for hours turning it around and loved how every move would create different colours and patterns. Later in life, when I was designing our garden, I wanted it to be like that kaleidoscope, with every month offering new patterns, shapes, colours, and textures. It can be challenging to create interest all year, but by reminding myself of that kaleidoscope I have successfully created a garden that gives me joy every single day.

Clockwise from top left: strawflowers (*Xerochrysum bracteatum*) continue to bloom now; the leaves of my staghorn sumach (*Rhus typhina*) are starting to fire up; dahlias provide home-grown bouquets throughout the autumn; forget-me-not seedlings found in the garden can be potted up now to bloom next spring.

IN SEASON

Autumn is a beautiful season, with lots of colourful leaves and flowers to enjoy, especially on sunny, mild days in October when a walk outside feels like a treat and reminds me that there are always treasures to be found in Nature.

WHAT'S IN BLOOM?

There is a surprising amount of colour in the garden in October, and even my lavender hedge has a few blooms left on it, luring the bees before winter sets in. Take your pick of these beautiful late-flowering and berried plants to create a garden rich in delights for you to enjoy, as well as birds and insects in search of an autumn feast.

Strawflower (*Xerochrysum bracteatum*)
I sow this upright annual each year, using the papery yellow, pink, and red flowers, which bloom through the summer and autumn, for dried flower arrangements to decorate my home during the winter months.

Japanese bugbane (*Actaea matsumurae*)
The perfect plant for a cool, shady place where the soil is not too dry, bugbane's elegant leaves and spikes of fluffy, bottle-brush white flowers bloom well into November in milder gardens.

Sedum (*Hylotelephium spectabile*)
Flowering up to the frosts, this easy-going plant is a must for any autumn garden. The flat-topped heads of tiny pink or white flowers draw in bees and butterflies, and then form rich dark brown seedheads that also provide food and habitats for wildlife over the winter. Very easy to propagate, just root the stems in water (see p.116) to fill your beds for free.

Crab apple (*Malus*)
If you're looking for a small deciduous tree with multiple seasons of interest, the crab apple is the perfect choice. Its colourful little fruits light up the garden at this time of year, and appear among bright autumn leaves in shades of yellow, orange, and red. The spring blossom is also loved by pollinators. My favourites include *Malus* 'Evereste', with its red-flushed fruits and *Malus* x *robusta* 'Red Sentinel', which produces dark red crabs.

Staghorn sumach (*Rhus typhina*)
I love this large deciduous shrub or small tree for its red stems and large divided leaves, which turn from green to red and orange in autumn, just as the cone-shaped heads of dark red berries appear. Its only drawback is that it produces lots of suckers, although you can dig these up and pot them on to make new plants.

Persicaria (*Bistorta amplexicaulis*)
These robust plants produce slim spikes of red, pink, or white flowers, which appear throughout late summer and autumn. It grows in sun or part shade and while it may prefer moisture-retentive soil, it is not too fussy and will cope with drier conditions.

Foxtail millet (*Setaria italica*)
This pretty annual grass is easy to grow from seed and produces a subtle leafy effect in

summer followed by millet-like seedheads in autumn. It is a great plant for cutting, and the birds love it, too.

Repeat-flowering roses
Many roses will continue to flower up to and even after the first frosts in autumn. Leave them unpruned now for hips over the winter.

Crimson flag (*Hesperantha coccinea*)
Not grown as frequently as they deserve, these autumn-flowering bulbs produce elegant spikes of lily-like flowers. Despite the common name, not all are crimson, with white and pink varieties offering more subtle colour ranges. Plant the bulbs in spring in well-drained soil or pots of peat-free compost, in full sun.

Moor grass (*Molinia caerulea*)
The beautiful native grass produces mounds of green leaves which turn yellow in autumn, while the bead-like seedheads also have impact in October before it dies down.

MINDFUL MOMENT

Rich with fiery leaf colours, October is a month to be enjoyed to the full. The garden offers so much happiness at this time of the year. The colours usually look more vibrant under the slanting sun, and the air is fresh and clear. I pick some bare stems and flowers for drying from my cutting garden – statice, strawflowers, and grasses – to make my annual autumn wreath and, in this way, celebrate the previous season's warmth that helped them to develop. It's so lovely to get creative with the precious materials that my garden delivers, and to express my gratitude for all the beauty I'm surrounded by. Take a moment to stop and applaud Nature and thank your garden for this year's show.

SPOTTING WILDLIFE

Wasp
The much-maligned wasp is, in fact, one of the gardeners' greatest allies, helping to keep pest populations in check and creating a healthy ecosystem, so before you dismiss these creatures as villains, I urge you to think again.

There are over 7,000 species of wasp living in the UK, and most are parasitoids. The larvae of these tiny wasps eat live insects or spiders, and some of the prey they consume eat our plants, so in this way they help to reduce the numbers of unwanted insects in the garden.

The larger black and yellow striped variety with the nasty sting also has its benefits, catching vast numbers of caterpillars and aphids to feed to their young. While the adults also indulge on our ice creams and jam sandwiches, in the wild, their sugar-rich diet consists of flower nectar and honeydew produced by aphids, as well as the sugary liquid released by their own larvae. As they visit flowers, they also help to pollinate them, just like bees. These social wasps also build the most spectacular papery nests, which you may find in wall and roof cavities, bird boxes, sheds or garages, after they have died off in autumn.

Right While wandering around the garden admiring the flowers, I'm always looking for ripe seedheads to harvest and sow for next year's show.

Clockwise from top left: take semi-ripe cuttings of perennials such as sedums; sow musk mallow (*Malva moschata*) for tall stems of pink summer flowers, loved by bees and butterflies; harvest pears while still hard, as they will ripen once picked; sow candytuft (*Iberis umbellata*) now indoors for beautiful flowers next year.

WHAT TO GROW IN OCTOBER

October is a good month for planting spring bulbs and hardy shrubs and perennials, and for dividing perennials where clumps have outgrown their allocated space.

SOW NOW INDOORS

Flowers

As well as planting bulbs outside for spring flowers next year, I also pot up some hyacinths and narcissus for indoor winter displays. Buy prepared hyacinth bulbs and narcissus bulbs now and plant them in bowls of bulb fibre with the pointed tips just showing above the surface. Keep the hyacinths in a cool, dark place at about 10°C (50°F) until you see shoots emerge, then bring them into a warm, light area. You can place the narcissus directly in the light to promote new growth. Good choices for winter cheer include: *Narcissus papyraceus* 'Ziva'; *N.* 'Erlicheer'; and *N.* 'Cragford'.

Other flowers to sow indoors in pots now include:

- Candytuft (*Iberis umbellata*)
- Sweet peas (*Lathyrus odoratus*) – sow in toilet rolls, to accommodate the long roots
- French lavender (*Lavandula stoechas*)
- Rose campion (*Lychnis coronaria*)

Herbs

Herb seeds such as basil, dill, chives, and parsley can be sown now and grown on a windowsill in late autumn and winter.

SOW OUTSIDE

Use mild days to sow flowers for next year's display. I use harvested seed from my plants and just scatter them about the beds to see what pops up.

Flowers

- Granny's bonnets (*Aquilegia*)
- Knapweed (*Centaurea nigra*)
- California poppies (*Eschscholzia californica*)
- Musk mallow (*Malva moschata*)

CUTTINGS TO TAKE

Semi-ripe cuttings

Take cuttings from currants and gooseberries, as well as non-flowering shrubs such as lavender and hydrangeas and perennial plants (see p.100). I keep my cuttings in pots outside in a sheltered area over winter.

Divide congested plants

Use this time to divide any plants that have outgrown their allotted space or performed poorly this year.

PLANT NOW

Self-sown seedlings

These free plants often appear in October and can be potted up and nurtured indoors or, if they are hardy, directly in beds (see p.25).

Autumn to spring flowers

Decorate sheltered areas of the garden, such as beside a front or back door, or a sunny wall, with pots of bellis daisies (*Bellis perennis*), primulas, and violas, including winter pansies.

Plant wallflowers

These biennials, which you sowed earlier in the year, can be planted in between your bulbs. You can also buy inexpensive bare-root plants now if you didn't have time to sow any.

Spring bulbs

Continue to plant spring bulbs in containers and the ground.

Plant perennials

Any hardy types taken from cuttings earlier in the year that have a robust root system can be planted now. Also plant new plants, which will settle in well now, and take cuttings from them when they are in full growth next year.

MAINTENANCE MUSTS

Harvest apples and pears

My old tree produces an abundance of delicious apples at this time of year, some of which I make into compote and jam, and store the rest to eat later. When storing apples and pears, place the fruits on slatted shelves in a cool, frost-free garage or room, so that air can circulate around them. Keep checking them and remove any that show signs of disease.

Also harvest the little crab apples from *Malus* trees and make them into jellies. I use the stems for autumn decorations, too.

Finish collecting seed

Most plants have set seed by now, except for the autumn-flowering perennials, so make sure you harvest what you need, and then leave the rest for the birds to enjoy over the coming weeks and months.

Final mow of the year

Grass growth slows as the temperatures fall, so give your lawn a final trim to keep it looking neat over winter. Leave some areas of longer grass for wildlife cover and habitats.

Bring in tender plants

Before the first frosts kill them, bring your tender plants inside, or, if plants are too big to move, take cuttings early in the month. In mild areas, you may be able to leave dahlias outside, covering the root area with a thick layer of chipped bark or compost for insulation.

Cut back perennials that have died down

Chop up the pruned stems and use as a mulch (see p.130) or leave perennials and ornamental grasses with structural seedheads standing throughout autumn and winter for the birds to enjoy and insects to shelter in.

ANYA'S MONEY-SAVING TIP

I like to visit gardens in autumn to see what plants others are growing at this time of year that inspire me. Local private gardens are often also good sources of inexpensive young plants that the owners have grown themselves from cuttings or seed. Or you could ask if they mind you taking a small stem of a plant you admire – most gardeners are very generous.

Clockwise from top left: the papery seedheads of honesty help to decorate the garden in winter, but I also take a few seeds to sow now; plant out wallflowers to bloom next spring; continue to plant daffodils and other bulbs; use crab apples to make jellies or add the stems of colourful fruits to your autumn indoor decorations.

ANYA'S PROPAGATION MASTERCLASS

DIVIDING PERENNIALS

Increasing your stocks of perennial plants by dividing them in autumn or early spring takes just a few minutes and all you need is a fork and spade to complete the task.

As perennial plants mature, most will start to spread, sending out roots underground from which new shoots will appear. Over time, they can extend beyond their allotted space and start to encroach on their neighbours. Congested clumps may also not flower well or their vigour can be reduced. Dividing them will help to reinvigorate flagging plants, while creating smaller clumps of fresh growth to transplant elsewhere in the garden, or to pot up and use in container displays.

You will need

Garden spade
Garden forks
Clean, sharp knife (optional)

1. Dig around a clump with a garden spade or fork, working it beneath the root ball so that you can gently lift the plant out of the soil.

2. Look for a natural gap between the stems and plunge your spade into it, cutting the clump into two or more sections. Plants with small, fibrous roots, such as heucheras and epimediums, can pulled apart with your hands, or, if the root ball is very congested, use a sharp knife to divide it up. For large plants, try plunging two forks back to back into the centre of the clump, and lever the root ball apart by pushing down on them.

3. Dig holes in other areas of the garden to accommodate your divided sections. Check each new clump, cutting out any dead or diseased roots and old stems, then planting them at the same depth they were growing at originally. Cut back the leaves of plants with large foliage after planting.

4. Water in well and wait for new growth to emerge in spring.

Aftercare

Water your divisions during dry spells in autumn and the following spring until the roots are well-established.

When to divide perennials

Divide summer-flowering plants in spring or autumn. Divide Mediterranean plants or those that are not fully hardy in the spring. Irises that grow from rhizomes (knobbly underground stems) are best divided in the summer, after flowering.

Suitable plants

Any perennial or bulbous plants, including agapanthus, anemones, asters, delphiniums, *Epimedium*, euphorbias, hardy geraniums (shown here), hosta, primulas, salvias, sedums, verbenas, and *Veronicastrum*.

1.
2.
3.
4.

PROJECT OF THE MONTH

GROW ROSES FROM A BOUQUET

Receiving a bouquet of roses from a loved one is always a special occasion and you can extend your enjoyment of these beautiful flowers and remember their gift by taking cuttings from the stems to grow new plants.

Roses that I have grown from cuttings help me to remember the happy times in my life and loved ones who have passed away, and because they have such deep significance for me, I have also grown plants from the bouquets I've been given. You may also like to propagate roses you received on a special occasion, such as a wedding, anniversary, or the birth of a baby, that have special meaning for you.

Bouquet cuttings are best from homegrown roses raised outside. (Note: it's illegal in the UK to propagate a patented rose variety.) Look for stems that don't bend too easily and have a few healthy leaves. Roses are more likely to root in the warmer months between early spring and early autumn, but you can try this method at any time of the year. After all, you have nothing to lose.

You will need

Sharp, clean secateurs and a knife
Seed tray or recycled food tray
Peat-free multipurpose compost mixed with some horticultural grit

1. Keep the roses in water until the flowers die. Then cut the stems into 7.5cm (3in) lengths, each with a leaf node (see p.86) in the centre.

2. Using a sharp knife, make a slanting cut on the opposite side of the stem to the leaf – do not cut through the whole stem.

3. Fill a tray with peat-free compost mixed with some horticultural grit and place the cut side of the stem on the surface – it is from this point that roots will form. You should be able to fit a few cuttings into one tray. Water the cuttings and keep in a sheltered area of the garden, out of direct sun, where the compost will not become waterlogged when it rains. In winter, keep them outside in a sheltered area.

4. Your cuttings should root between 4 and 8 weeks later, depending on the season. You can then pot them up into containers of their own to grow on. The plants should produce some blooms about a year or two later. While the roses your cuttings produce will be genetically identical to the parent plants, the shrubs may not be as vigorous since most roses are grown on different root stocks. The flowers may also be smaller than those in the original bouquet but you can still create something meaningful and enjoy with beautiful plants and flowers.

1.

2.

3.

4.

CHAPTER 11

NOVEMBER

November may feel like the end of the show, as the days get shorter and the frost takes the last blooms, but even late autumn can offer reasons to be grateful for our gardens. I often walk around mine early in the morning to admire all the beautiful seedheads, rustling grasses, and interesting shapes. The last strawflowers are usually hanging on in my cutting garden, ready for their final harvest, while the shrubs and trees look like they're on fire before the foliage falls, leaving a carpet of gold, amber, and red on the ground.

The short days can also be a blessing, giving us time to reflect on the previous seasons. It feels good to slow down and take stock. And not everything is dying – I am always excited to see new growth on the sweet pea seedlings I sowed last month, along with the shoots on cuttings I took at the end of the summer. For me, this is the joy of gardening. When one season comes to an end, another begins and the cycle of life is right there in front of your eyes.

IN SEASON

Late autumn delivers a feast of fiery leaf colours and bright berries, while a few perennials also put on a performance when others are taking their annual rest below ground.

WHAT'S IN BLOOM?

The hydrangea flowers may be over but their beautiful seedheads continue to decorate the garden, while many trees and shrubs are laced with bright berries, the jewel-like fruits often appearing fleetingly before the birds strip the stems. Others performing now include:

Stinking iris (*Iris foetidissima*)
Despite its unfortunate name, this shade-tolerant iris does not smell bad, unless you crush the leaves, and has much to admire, with June flowers followed by orange berries now.

Guelder rose (*Viburnum opulus*)
This beautiful native shrub produces large white flowerheads in spring, followed in autumn by clusters of bright red berries, loved by birds. 'Xanthocarpum' has golden berries in autumn and is a little smaller than the species.

Japanese saxifrage (*Saxifraga fortunei*)
Flowering in November, this saxifrage is a great addition to any garden, its small blooms exploding like tiny fireworks when they appear. The scalloped leaves also have decorative value earlier in the year.

Evergreen ferns
Hardy ferns that retain their leaves over winter offer colour and texture in autumn and winter. My favourites include the hard shield fern (*Polystichum aculeatum*), with its bright green triangular feathery fronds, and the hart's tongue fern (*Asplenium scolopendrium*), which produces wavy-edged glossy green leaves.

MINDFUL MOMENT

Very soon, flowers in my cutting garden will be taken by the frost and November offers my last chance to arrange them, but I'm not sad. I'm very mindful about it. Nature is a great life coach, teaching us acceptance and patience, and giving us hope for tomorrow. When I grow flowers to pick, I express my gratitude for each stem. You may also find it mindful to look at the delicate petals, feel the textures, and inhale the scents of those in your garden, and don't feel unhappy when they go – they'll soon return to bring more happiness.

SPOTTING WILDLIFE

Earthworm
The humble earthworm is often taken for granted, wiggling its way through the soil, unseen and unheard. However, this little invertebrate is perhaps the most important creature in your garden, helping to recycle dead leaves, stems, and flowers, pulling them down into the soil before consuming them, then delivering the nutrients they hold back to feed plant roots. Worm droppings or casts also have another critical role to play, helping to improve the structure of the soil, while the tunnels these creatures make also allow air and water to reach your plants' roots.

Clockwise from top left: leave the dried flowerheads of hydrangeas standing to provide habitats for overwintering insects; the stinking iris has beautiful orange berries in autumn; evergreen ferns create a textured frill around shrubs and trees; *Viburnum opulus* 'Xanthocarpum' produces golden berries in autumn.

Clockwise from top left: pots of alpines can be placed in a sheltered spot outside over winter or in an unheated greenhouse; planting tulip bulbs in November helps to prevent viruses and diseases; plant winter pansies in pots to decorate your patio; prune roses lightly now to prevent wind rocking the stems and tearing their roots.

WHAT TO GROW IN NOVEMBER

November is a great time to plant trees and shrubs, which are available now as bare-root plants. Usually much cheaper than pot-grown plants, these are grown in a field, then dug up when they are dormant and sent directly to your door.

SOW INDOORS NOW

Flowers

It's not too late to sow sweet peas at the beginning of the month, but the colder temperatures and low light levels mean germination rates may be slower at this time of year. Also keep plants sown earlier on a bright windowsill where they will receive the maximum sunlight to fuel their growth.

PROPAGATE NOW

Divide shop-bought basil

Pots of basil are usually root-bound and only lasts a few days, so divide it into three plants and pot them up into some fresh compost.

Take hardwood cuttings

Use this simple method to increase shrubs and roses (see p.156).

PLANT NOW

Tulip bulbs

Planting tulips in late autumn is thought to help to slow the spread of viruses and tulip fire, which is prevalent earlier in the season.

Bare-root trees, shrubs, and hedges

November is the best time to plant trees, especially for money-saving gardeners, since you can now pick up cheaper bare-root plants. Shrubs and hedging, as well as some perennials, are also available as bare-root plants now (see p.13 and p.21 for planting tips).

MAINTENANCE MUSTS

Clear leaves from small ponds

Autumn leaves can pollute the water in small ponds and features when they biodegrade and release their nutrients, so scoop them out with a fishing net or cover the surface with netting, making sure amphibians and other wildlife can still get in and out.

Prune roses to prevent windrock

Take stems down by about a third to prevent wind rocking the plants and tearing the roots. You can then prune them again in late winter or early spring to promote new growth.

Plant out winter pots

Using bedding sown earlier in the year and shrub cuttings, I like to create some pretty pots that I can see from my windows in winter.

Mulch beds and borders

Covering the soil with a 5cm (2in) layer of homemade compost now can help to protect it from erosion in winter, while a deep mulch over the roots of slightly tender plants can also help them to survive the colder months.

ANYA'S PROPAGATION MASTERCLASS

HARDWOOD CUTTINGS

Many of the shrubs, climbers and roses in my garden were grown from hardwood cuttings taken in autumn and winter. Your new rooted plants will be ready to pot on or plant out the following year.

This method of propagation is a great way to increase deciduous shrubs, climbers, and trees, and I also find it very mindful to do when the rest of the garden is quiet and taking a rest.

It takes just a few minutes to snip off some stems to propagate in the ground outside, and because they remain tucked up in the soil, they need almost no aftercare, bar checking that the soil doesn't dry out in summer. I also pop some dogwood (*Cornus*) stems into a large urn filled with soil-based compost, since I have so little space for these cuttings in my borders.

You will need

Clean, sharp knife or secateurs
Sand or grit
Large container of loam-based compost (optional)

1. After leaf fall, select a healthy stem that grew this year and remove the soft tips. Trim into sections 15–45cm (6–18in) long, cutting above a bud – make a sloping cut so that water will drain away from the bud and to remind you which end should be planted at the top. Then cut straight across at the base, below a bud or pair of buds.

2. Dig a trench in a quiet, sheltered area of the garden where the soil will not become waterlogged overwinter, or make holes in a large container if that is easier. Add a layer of sand or grit to the base of the trench or holes, then insert the cuttings, with two-thirds of the stems below the surface and a few buds above it to allow the plant to grow in spring.

3. The following spring, you will see shoots developing on your cuttings, and this is a sign that new roots may also be forming. Leave the cuttings in place until the following autumn, when you can carefully dig them up.

4. Replant the cuttings where you want them to grow, either in the ground or in a pot of loam-based compost or garden soil. Water in well, and keep watered during the first couple of seasons until the roots are well-established.

When to take cuttings

Any time from mid-autumn until late winter, after the leaves have fallen.

Suitable plants

- Roses, currant bushes, and most deciduous shrubs, including *Abelia*, *Deutzia*, the butterfly bush (*Buddleja*), dogwoods (*Cornus*), forsythia, and viburnums
- Climbers, such as honeysuckle (*Lonicera*), jasmine, vines (*Vitis*), and *Parthenocissus*
- Trees, including planes (*Platanus*) and willows (*Salix*)

1.

2.

3.

4.

PROJECT OF THE MONTH

POTTED BULBS AND SUCCULENTS

The perfect seasonal gift, you can make this pretty planter filled with hyacinths that will bloom next spring and some little echeverias to admire through the winter months, nestling in a bed of moss.

I love making seasonal gifts for my family and friends, and for myself, to bring some natural beauty into the house during the dark days of winter. I am using prepared hyacinth bulbs here, which you can buy now. These have been heat-treated to prompt them to flower earlier than normal – they bloom in April when planted outside in the garden.

You will need

Multipurpose compost mixed with some horticultural grit
Pot or planter with drainage holes in the base
Prepared hyacinth bulbs
Echeveria plants (or *Sempervivum*)
Moss
Pine cones (optional)

1. Add the compost/grit mixture to your container, filling it up to about 5cm (2in) below the rim. Then plant the hyacinths, so the pointed tips are just showing above the surface of the compost.

2. Plant your echeverias between the bulbs. I have grown these from offsets (see p.54). You could also use hardier houseleeks (*Sempervivum*), which can be placed outside in a sheltered spot in colder areas.

3. Place the moss between the plants. I raked up mine from the lawn, but you can buy moss in packs at the garden centre. Pines cones add a little extra decorative touch.

4. Place the pot in a sunny, sheltered area outside – the echeverias are not fully hardy but will be fine outside in milder regions. Alternatively, set the pot in an unheated greenhouse. Water sparingly about once every two or three weeks.

Aftercare

When the hyacinths start to sprout, I prop up the stems with a few beech hedge or dogwood prunings (other woody twigs will also work) so that their stems do not overshadow the echeverias. Alternatively, you can repot the succulents into containers of their own at this stage, if that's easier.

After the bulbs have flowered, plant them outside in the garden, where they will bloom the following year in April. Repot the succulents in spring and summer containers.

1.
2.
3.
4.

CHAPTER 12

DECEMBER

Creating a garden is like writing a theatre script. You want the audience to be entertained until the end, and winter in the garden represents the final curtain call. Full of interesting textures and shapes, it can look truly magical in the depths of December, when caught by a shaft of late afternoon sun or dusted with snow and ice.

Many people struggle during the cold dark days of winter, even during the festive period, but the garden can be very healing at this time of year. I often go outside on a foggy morning and look at the gleaming mahogony bark on *Prunus serrula* and peeling stems of *Acer griseum*, known as the paperbark maple, to lift my spirits.

December is also a good time to visit winter gardens that are open at this time of year. You will find plenty of inspiration for your own garden, and a walk in the fresh air, perhaps with a hot chocolate to end the day, is guaranteed to make you smile.

IN SEASON

Winter is here and many plants have retreated underground, but there are still lots to enjoy. I love the deciduous trees and shrubs, stripped of leaves to reveal their intricate network of branches, or the hollies decked with bright berries.

WHAT'S IN BLOOM?

As the winter festivities approach, think about planting traditional plants such as holly and ivy to make your own wreaths and decorations.

Wintersweet (*Chimonanthus praecox*)
This medium-sized shrub is a must for winter interest. The highly scented waxy flowers are perfect for indoor arrangements, and it's easy to propagate by layering (see p.26).

Holly (*Ilex aquifolium*)
For berries, you need to choose a female plant such as *Ilex aquifolium* 'Handsworth New Silver' or 'Madame Briot', which both have red fruits and variegated leaves.

Ivy (*Hedera helix*)
Providing food and habitats for wildlife, I also use the twining stems to make wreaths and garlands for the house in December. Many forms are very vigorous and will need to be pruned back in winter to keep them in check.

Winter honeysuckle (*Lonicera fragrantissima*)
A beautiful addition to the winter garden, this hardy shrub produces an abundance of scented creamy-white flowers from December to March on almost leafless branches, sometimes followed by red berries.

MINDFUL MOMENT

There's something so wonderful about just being. I am normally a very sociable person, filling my days in December with activities – meeting friends and family – but the winter festivities also offer time for reflection. It's great to take a break from work and just enjoy the simple pleasures of cooking, watching TV, spending time with my children, and going on woodland walks. Even a short walk in winter can be magical, revealing natural treasures – a shiny holly leaf, seedheads covered with frost, and the odd rose still in bloom, adding a little faded sparkle to the celebratory mood.

SPOTTING WILDLIFE

Robin
Adorning Christmas cards and gifts, the little robin is one of my favourite birds. One always accompanies me when I'm working in the garden, looking for any worms I may unearth.

Both males and females have red breasts and look almost identical, but their young are spotted and golden brown, developing their true colours a few months after fledging Despite their seemingly friendly manner, these birds are feisty and guard their territory aggressively, driving away intruders until the mating season arrives. They eat worms, seeds, fruits, insects and other invertebrates, so include a range of plants to lure them in.

Clockwise from top left: only female or self-fertile holly cultivars bear winter berries; wintersweet (*Chimonanthus praecox*) flowers from late December; plant ivy to support wildlife; include berried plants for robins but avoid feeders, which artifically increase bird numbers, leading to a fall in butterfly larvae populations in spring.

Clockwise from top left: leave the seedheads of perennials such as sea hollies (*Eryngium*) to decorate the winter garden and support wildlife; wait until after leaf-fall to take hardwood cuttings of shrubs such as *Euonymus alatus*; put up nest boxes to provide birds with shelter during the winter; leave rose hips for wildlife to enjoy.

WHAT TO GROW IN DECEMBER

Continue planting bare-root trees, shrubs, and hedges throughout December, when the weather allows, and take hardwood cuttings of roses, shrubs, and climbers.

SOW INDOORS

I don't sow any seeds during December, when the light is poor. Instead, I use the time to look through my seed catalogues and online for new plants I would like to grow next year.

PROPAGATE NOW

Hardwood cuttings

Continue to take cuttings of shrubs, climbers, and trees now (see p.156).

PLANT NOW

Continue planting spring bulbs

As long as the bulbs are firm and disease-free they will still flower next year if planted up to the middle of December.

Bare-root trees and shrubs

Continue to plant bare-root woody plants when the weather and soil conditions allow (see p.13). You can also use these plants or shrub cuttings taken last year to make an inexpensive hedge.

MAINTENANCE MUSTS

Break ice on ponds and water features

To ensure birds and other wildlife have water to drink during cold periods, place a tennis ball on the surface of your pond, then remove it when ice forms to create a hole, or pour on some warm water to melt it.

Prune apple and pear trees

I like to prune my old apple tree on a sunny day in December. First, I take out old, diseased, and dead stems, and then try to create an open framework so that light and air reach all areas of the tree. Don't overprune, though – remove no more than 10–20 per cent of the canopy each year. Taking out more growth will result in the formation of unproductive water shoots.

Prune acers, birches, and vines

Cut back these plants before Christmas to avoid the stems bleeding sap.

Prune autumn raspberries and currants

Autumn raspberries are very easy to prune this month by taking all the stems down to the ground. Currants can be pruned from the end of December to early spring before new growth starts to appear. Cut back the stems to create an open framework that will allow light and air to reach the fruits next year.

Install nesting boxes

Birds often use these boxes to shelter in during the winter and they may then nest in them later in spring. Buy different shaped boxes for different birds and check with an authoritative bird charity, such as the British Trust for Ornithology (BTO), for information on the best areas of the garden to install them.

PROJECT OF THE MONTH
VASE OF AMARYLLIS

The large, colourful flowers of amaryllis always cheer me up when they flower in late winter, and I have now learned how to keep the bulbs from year to year, saving me the cost of buying new ones.

Amaryllis (*Hippeastrum*) remind me of my home in Poland. My granny and mum always grew them and the emerging flower stems were always a topic of discussion around the kitchen table. When the amaryllis finally bloomed, lighting up our home like floral fireworks, the whole family was informed, and everyone would come to admire them.

These bulbs are easy to grow, and you could try my favourites, 'Apple Blossom', with pale pink-infused white flowers, and 'Magic Green', which sports white flowers with green and dark red markings.

You will need

Tall vase or container
Horticultural grit
Bulb fibre or peat-free multipurpose compost mixed with a handful of horticultural grit
Amaryllis bulbs
Moss and pine cones (optional)

1. If your container doesn't have drainage holes, add a 2.5cm (1in) layer of grit, then a layer of bulb fibre or compost mix on top. Place the bulbs, pointed end facing up, on top, making sure that they are not touching.

2. Add some more compost, so that the top third of the bulbs are above the surface. I also add some moss, raked from the lawn, to keep the bulbs slightly moist, and a few pine cones for decoration. Water very lightly around the bulbs to moisten the compost.

3. Place the vase in a warm, bright spot, and keep the compost damp but not wet. Stems will soon emerge from the bulbs, followed in a few weeks by the large, lily-like flowers.

4. To lengthen flowering time, once the flowers start to show, move the container to a cooler place. Turning it now and again to prevent the stems from bending towards the light. Do not allow your plants to dry out.

ANYA'S TOP MONEY-SAVING TIP

Rather than discarding the bulbs after flowering, keep them for the following year. Remove the old flower stems after they have faded, and continue to water the bulbs, adding a little all-purpose fertilizer every two weeks in spring and summer, and leaving the new foliage to grow. When the leaves turn in early autumn, cut them back to about 5cm (2in) from the top of the bulb. Remove the bulbs from the compost, and place them in a cool, dark place for a minimum of six weeks, then replant them as shown here.

1.

2.

3.

4.

PLANT PROPAGATION GUIDE

This easy-to-use guide shows you the best propagation methods and timings for the plants in the book, together with a selection of other popular species.

The propagation techniques are described in more detail on the pages below:

PLANT	Propagation method	Months											
		J	F	M	A	M	J	J	A	S	O	N	D
Abelia	layering												
	semi-ripe cuttings												
	softwood cutting												
Acer	harwood cuttings												
Achillea	divison												
	seed sown under cover												
	semi-ripe cuttings												
Aconitum carmichaelii	division												
Actaea simplex	division												
Agapanthus	division												
	seeds sown under cover												
Agastache	basal cuttings												
	division												
	seeds sown under cover												
Agave	offsets												
Akebia quinata	layering												
	seeds sown under cover												
	softwood cuttings												
Alcea rosea	seeds sown directly												
Alchemilla mollis	division												
	seeds sown under cover												
Allium	offsets												
Aloe vera	offsets												
Alstroemeria	division												
Ammi majus	seeds sown directly												
	seeds sown under cover												
Anemone (Japanese types)	division												
	seeds sown under cover												
Anethum graveolens (dill)	seeds sown directly												
	seeds sown under cover												

PLANT	Propagation method	Months											
		J	F	M	A	M	J	J	A	S	O	N	D
Antirrhinum	seeds sown directly					■	■		■	■	■		
	seeds sown under cover		■	■						■	■		
Aquilegia	seeds sown directly		■	■	■	■				■	■		
	seeds sown under cover	■	■	■	■	■				■	■		
Arabis	cuttings of rosettes					■	■	■	■				
Argyranthemum	semi-ripe cuttings							■	■	■	■		
	softwood cuttings			■	■	■							
Artemisia	basal cuttings			■	■	■							
	division			■	■	■				■	■		
	seeds sown under cover	■	■	■	■				■	■	■	■	
Asters (including *Symphyotrichum*)	seeds sown under cover		■	■	■								
	semi-ripe cuttings								■	■	■		
	softwood cuttings			■	■	■							
Astilbe	division			■	■	■				■	■		
	seeds sown directly					■	■						
	seeds sown under cover		■	■	■	■	■			■	■		
Astrantia	division			■	■	■				■	■		
	seeds sown directly									■	■	■	■
	seeds sown under cover	■	■	■						■	■	■	■
Athyrium	division			■	■	■							
Aubrieta	seeds sown directly		■	■	■	■	■			■	■		
	seeds sown under cover	■	■	■	■	■	■						■
	softwood cuttings after flowering			■	■	■							
Aucuba	layering			■	■	■							
	semi-ripe cuttings								■	■	■		
Bacopa	layering		■	■	■								
	semi-ripe cutting								■	■	■		
	softwood cuttings			■	■	■							
Begonia	basal cuttings				■	■	■						
	division			■	■	■				■	■	■	
	seeds sown under cover	■	■	■	■								

PLANT	Propagation method	Months											
		J	F	M	A	M	J	J	A	S	O	N	D
Bellis perennis	division								■	■	■		
	seeds sown directly				■	■	■	■	■	■	■		
	seeds sown under cover	■	■	■									
Berberis	heel cuttings								■	■	■		
	seeds sown under cover			■	■						■	■	
	semi-ripe cuttings								■	■	■		
Bergenia	division			■	■	■				■	■	■	
	seeds sown directly	■	■	■	■								
	seeds sown under cover	■	■	■									
Betula pendula	softwood cuttings			■	■	■							
Bistorta amplexicaulis (Persicaria)	division			■	■	■				■	■		
	softwood cuttings			■	■	■							
Briza maxima	seeds sown directly				■	■			■	■			
	seeds sown under cover		■	■									
Brunnera macrophylla	division			■	■	■				■	■	■	
Buddleja	hardwood cuttings	■	■									■	■
	semi-ripe cuttings									■	■		
	softwood cuttings			■	■	■							
Calamagrostis x *acutiflora*	division			■	■								
Calendula	seeds sown directly				■	■							
	seeds sown under cover			■	■					■	■	■	
Camellia	layering			■	■	■							
	hardwood cuttings	■	■									■	■
	semi-ripe cuttings								■	■	■		
Campanula	division			■	■	■				■	■	■	
	seeds sown under cover	■	■	■						■	■	■	■
Capsicum annuum (chilli)	seeds sown under cover		■	■									
	semi-ripe cuttings								■	■	■		
	softwood cuttings			■	■	■							

PLANT	Propagation method	Months											
		J	F	M	A	M	J	J	A	S	O	N	D
Caryopteris x *clandonensis*	semi-ripe cuttings									■	■	■	
	softwood cuttings			■	■	■			■	■	■		
Catananche caerulea	division			■	■	■							
	seeds sown directly			■	■	■							
	seeds sown under cover		■	■	■								
Ceanothus	softwood cuttings			■	■	■							
	semi-ripe cuttings								■	■	■		
Celosia argentea	seeds sown under cover		■	■									
Centaurea montana	division (runners)			■	■								
Centranthus ruber	heel cuttings			■	■								
	seeds sown directly				■	■	■						
	seeds sown under cover			■	■	■	■						
Cercis canadensis	softwood cuttings			■	■	■							
Cerinthe major	seeds sown directly					■	■						
	seeds sown under cover			■	■				■	■			
	softwood cuttings			■	■	■							
Chaenomeles	air layering			■	■	■				■	■	■	
	layering				■	■				■	■	■	
	semi-ripe cuttings							■	■	■			
Chimonanthus praecox	air layering							■	■	■			
	layering							■	■	■			
	softwood cuttings			■	■	■							
Chionodoxa	offsets						■	■	■				
Chlorophytum comosum	offsets	■	■	■	■	■	■	■	■	■	■	■	■
Choisya	air layering			■	■	■							
	layering			■	■	■							
	semi-ripe cuttings								■	■	■		
Chrysanthemum	basal cuttings			■	■	■							
	seeds sown directly				■	■	■						
	seeds sown under cover			■	■								
	semi-ripe cuttings								■	■			

PLANT	Propagation method	J	F	M	A	M	J	J	A	S	O	N	D
		Months											
Cirsium rivulare	division			■	■	■				■	■	■	
Clarkia	seeds sown directly			■	■								
	seeds sown under cover								■	■	■	■	
Cleome	seeds sown directly					■	■	■					
	seeds sown under cover	■	■	■									
Clematis	internodal cuttings							■	■	■			
	layering			■	■	■							
Cobaea scandens	internodal cuttings						■	■	■				
	seeds sown under cover	■	■	■									
Consolida regalis	seeds sown directly			■	■	■							
	seeds sown under cover								■	■	■		
Coreopsis	division			■	■								
	seeds sown under cover		■	■									
Cornus	layering			■	■	■				■	■	■	
	hardwood cuttings	■	■									■	■
	semi-ripe cuttings								■	■	■		
Corydalis	division									■	■	■	
Cosmos	seeds sown directly				■	■							
	seeds sown under cover			■	■	■							
	softwood cuttings				■	■							
Cotinus	air layering			■	■	■				■	■	■	
	layering			■	■	■				■	■	■	
	hardwood cuttings	■	■									■	■
Cotoneaster	semi-ripe cuttings								■	■	■		
Coriandrum (coriander)	seeds sown directly			■	■	■	■		■	■			
	seeds sown under cover		■	■	■				■	■	■		
Crataegus	seeds sown directly								■	■	■		
Crocosmia	division of corms		■	■									
Crocus	division					■	■	■					
Cyclamen coum	division					■	■	■					

PLANT	Propagation method	Months											
		J	F	M	A	M	J	J	A	S	O	N	D
Cynara cardunculus	division			■	■								
	seeds sown under cover		■	■									
Dahlia	basal cuttings		■	■	■								
	division		■	■	■						■	■	
	softwood cuttings		■	■	■								
Daphne	layering			■	■	■							
	semi-ripe heel cuttings								■	■	■		
Delphinium	basal cuttings			■	■	■							
	heel cuttings			■	■	■							
	seeds sown under cover	■	■	■	■					■	■	■	■
Deutzia	semi-ripe cuttings								■	■	■		
	softwood cuttings			■	■	■							
	hardwood cuttings	■	■									■	■
Dianthus (perennial types)	seeds sown under cover		■	■	■								
	semi-ripe cuttings								■	■	■		
	softwood cuttings			■	■	■							
Dianthus barbatus	seeds sown directly					■	■	■					
Diascia	seeds sown under cover			■	■								
	softwood cuttings			■	■	■							
	semi-ripe cuttings								■	■	■		
Dicentra eximia	division			■	■	■							
Dictamnus	seeds sown under cover	■	■	■									■
Digitalis	seeds sown directly					■	■						
	seeds sown under cover		■	■	■								
Dipsacus fullonum	seeds sown directly				■	■							
Echeveria	seeds sown under cover	■	■	■	■	■	■	■	■	■	■	■	■
	offsets			■	■								
Echinacea	division			■	■					■	■	■	
	seeds sown under cover		■	■	■	■			■	■	■	■	
Echinops	division			■	■	■				■	■	■	
	seeds sown under cover		■	■	■	■			■	■	■	■	

PLANT	Propagation method	Months											
		J	F	M	A	M	J	J	A	S	O	N	D
Elaeagnus	seeds sown under cover	■	■	■	■	■	■	■	■	■	■	■	■
	semi-ripe cuttings								■	■	■		
Epimedium	division			■	■	■				■	■	■	
	semi-ripe cuttings								■	■	■		
	softwood cuttings			■	■	■							
Eranthis hyemalis	division					■	■						
	seeds sown under cover		■	■	■	■	■			■	■	■	
Erigeron karvinskianus	division			■	■	■							
	seeds sown directly				■	■	■						
	seeds sown under cover		■	■	■	■	■	■					
	softwood cuttings			■	■	■							
Erysimum cheiri	seeds sown directly								■	■			
	seeds sown under cover	■	■	■	■								
Eryngium	division			■	■					■	■	■	
	seeds sown under cover	■	■	■	■	■	■	■	■	■	■	■	■
Eschscholzia californica	seeds sown directly			■	■	■							
	division after flowering					■	■						
Euonymus fortunei	semi-ripe cuttings								■	■	■		
Euphorbia	seeds sown directly									■	■	■	■
	seeds sown under cover		■	■	■	■	■	■					
	softwood cuttings			■	■	■							
Fagus	seeds sown directly		■							■	■	■	
Fatsia	semi-ripe cuttings								■	■	■		
	seeds sown under cover	■	■	■	■	■	■	■	■	■	■	■	■
Ferns	division			■	■								
Foeniculum vulgare	seeds sown directly			■	■	■							
Forsythia	hardwood cuttings	■	■									■	■
	layering			■	■	■				■	■	■	
	semi-ripe cuttings								■	■	■		

PLANT	Propagation method	Months											
		J	F	M	A	M	J	J	A	S	O	N	D
Fuchsia	layering			■	■	■							
	semi-ripe cuttings								■	■	■		
	softwood cuttings			■	■	■							
Galanthus	division after flowering			■	■	■							
Geranium	division			■	■	■			■	■	■		
Geum	division			■	■	■				■	■	■	
	seeds sown under cover		■	■	■	■	■	■					
Gillenia trifoliata	division			■	■	■				■	■	■	
	seeds sown under cover	■	■	■	■	■	■	■	■	■	■	■	■
Hakonechloa macra	division			■	■	■							
Hamamelis	layering			■	■	■							
Hebe	semi-ripe cuttings								■	■	■		
	softwood cuttings			■	■	■							
Hedera helix	softwood cuttings	■	■	■	■	■	■	■	■	■	■	■	■
	layering	■	■	■	■	■	■	■	■	■	■	■	■
Helenium	division			■	■	■				■	■		
	semi-ripe cuttings								■	■	■		
	softwood cuttings			■	■	■							
Helianthus	seeds sown directly				■	■							
	seeds sown under cover			■	■								
Heliotropium arborescens	seeds sown under cover		■	■	■								
Helleborus	division			■	■	■				■	■	■	
	seeds sown directly			■	■	■							
	seeds sown under cover	■	■	■	■	■	■	■	■	■	■	■	■
Hemerocallis	division			■	■	■				■	■	■	
Hesperantha coccinea	division			■	■	■							
Hesperis matronalis	seeds sown directly					■	■	■					
	seeds sown under cover		■	■	■								
Heuchera	division			■	■	■							
	offsets			■	■	■							

PLANT	Propagation method	J	F	M	A	M	J	J	A	S	O	N	D
Heuchera	seeds sown under cover	■	■	■	■	■	■	■	■	■	■	■	■
Hippeastrum	offsets	■	■	■									
Hordeum jubatum	seeds sown directly				■	■	■						
	seeds sown under cover		■	■	■								
Hosta	division		■	■						■	■	■	
Hydrangea	hardwood cuttings	■	■									■	■
	semi-ripe cuttings								■	■	■		
	softwood cuttings			■	■	■							
Hydrangea petiolaris (climbing)	layering			■	■	■				■	■	■	
	softwood cuttings			■	■	■							
Hylotelephium (syn. *Sedum*)	division			■	■	■				■	■	■	
	semi-ripe cuttings								■	■	■		
	softwood cuttings			■	■	■							
Ilex	air layering			■	■	■							
	layering			■	■	■							
	semi-ripe cuttings								■	■	■		
Ipomoea	seeds sown directly					■	■						
	seeds sown under cover		■	■	■								
Iris	division after flowering					■	■						
Jasminum	layering			■	■	■							
	semi-ripe cuttings								■	■	■		
Knautia macedonica	division			■	■	■				■	■	■	
	seeds sown directly									■	■	■	
	seeds sown under cover		■	■	■	■	■						
Kniphofia	division			■	■	■	■	■	■				
	offsets						■	■	■				
Lagurus ovatus	seeds sown directly			■	■								
Lamprocapnos spectabilis (syn. *Dicentra*)	division			■	■	■							

PLANT	Propagation method	Months											
		J	F	M	A	M	J	J	A	S	O	N	D
Lathyrus odoratus	seeds sown directly			■	■	■							
	seeds sown under cover	■	■	■	■						■	■	
	softwood cuttings			■	■	■							
Laurus nobilis (bay)	layering			■	■	■							
	semi-ripe cuttings								■	■	■		
Lavandula	hardwood cuttings	■	■								■	■	■
	semi-ripe cuttings							■	■	■			
	softwood cuttings			■	■	■	■						
Leucanthemum × *superbum*	division			■	■	■				■	■	■	
Liatris spicata	division			■	■								
	seeds sown directly			■	■	■							
	seeds sown under cover		■	■	■	■			■	■	■	■	
Limnanthes	seeds sown directly			■	■	■							
	seeds sown under cover		■	■									
Limonium	seeds sown under cover		■	■	■								
Linaria x *purpurea*	division			■	■	■							
	seeds sown directly				■	■				■			
	seeds sown under cover		■	■	■				■	■			
	softwood cuttings			■	■	■							
Liriope muscari	division			■	■	■							
Lobelia	division			■	■	■							
	seeds sown under cover	■	■	■	■								
	softwood cuttings			■	■	■							
Lonicera (climbers)	seeds sown under cover	■	■	■	■	■	■	■	■	■	■	■	■
	softwood cuttings			■	■	■							
	semi-ripe cuttings								■	■	■		
Lunaria annua	seeds sown directly			■	■	■	■	■	■	■	■		
	seeds sown under cover			■	■	■							

PLANT	Propagation method	J	F	M	A	M	J	J	A	S	O	N	D
Lupinus	basal cuttings			■	■	■							
	heel cuttings			■	■	■							
	seeds sown directly				■	■							
	seeds sown under cover			■	■	■							
Lysimachia atropurpurea	division			■	■	■							
	seeds sown under cover		■	■	■					■	■	■	
Lythrum salicaria	division			■	■	■							
	seeds sown directly				■	■	■	■	■	■	■		
	seeds sown under cover	■	■	■	■	■	■	■	■	■	■	■	■
Magnolia stellata	air layering			■	■	■							
	layering			■	■	■							
	semi-ripe cuttings								■	■	■		
Mahonia	semi-ripe cuttings								■	■	■		
Malva moschata	basal cuttings			■	■	■							
	seeds sown directly									■	■		
	seeds sown under cover		■	■	■	■	■						
Matthiola incana	seeds sown under cover		■	■									
Mentha suaveolens (mint)	division			■	■	■	■						
	softwood cuttings		■	■	■	■	■	■	■	■	■		
Miscanthus sinensis	division			■	■	■							
	seeds sown under cover		■	■	■								
Molinia	division			■	■	■							
	seeds sown under cover		■	■	■	■				■	■	■	
Monarda	seeds sown directly				■	■	■	■	■				
	seeds sown under cover		■	■	■	■							
	semi-ripe cuttings								■	■	■		
	softwood cuttings			■	■	■							
Myosotis sylvatica	seeds sown directly					■	■						
	seeds sown under cover					■	■			■			
	division				■	■	■						

PLANT	Propagation method	J	F	M	A	M	J	J	A	S	O	N	D
		Months											
Narcissus	offsets					■	■						
Nemesia	seeds sown under cover			■	■	■				■	■	■	
	softwood cuttings			■	■	■			■	■			
Nepeta	division			■	■	■				■	■	■	
	semi-ripe cuttings								■	■	■		
	softwood cuttings			■	■	■							
Nerine bowdenii	division				■	■	■						
Nicotiana	seeds sown under cover			■	■								
Nigella	seeds sown directly			■	■	■	■	■	■	■			
Oenothera lindheimeri (gaura)	seeds sown under cover	■	■	■	■	■							
	semi-ripe cuttings								■	■	■		
	softwood cuttings			■	■	■							
Origanum (oregano)	division			■	■	■				■	■	■	
	seeds sown under cover	■	■	■	■	■	■	■	■	■	■	■	■
	softwood cuttings			■	■	■							
Orlaya grandiflora	seeds sown directly				■	■				■	■		
	seeds sown under cover			■									
Ocimum basilicum (basil)	seeds sown under cover	■	■	■	■	■	■	■	■	■	■	■	■
Osteospermum	seeds sown under cover	■	■	■									
	semi-ripe cuttings								■	■	■		
	softwood cuttings			■	■	■							
Paeonia	division									■	■	■	
Paeonia × suffruticosa (tree peony)	air layering			■	■	■				■	■	■	
	layering			■	■	■				■	■	■	
	semi-ripe cuttings								■	■	■		
Panicum	division			■	■	■							
Papaver orientale	seeds sown directly					■	■	■	■	■			
	seed sown under cover		■	■	■					■	■	■	
Parthenocissus	layering	■	■	■	■	■	■	■	■	■	■	■	■
Passiflora	layering			■	■	■				■	■	■	
	seeds sown under cover	■	■	■	■	■	■	■	■	■	■	■	■
	semi-ripe cuttings								■	■	■		

PLANT	Propagation method	Months											
		J	F	M	A	M	J	J	A	S	O	N	D
Pelargonium	seeds sown under cover	■	■	■									
	semi-ripe cuttings								■	■	■		
	softwood cuttings			■	■	■							
Penstemon	seeds sown under cover		■	■						■	■	■	
	semi-ripe cuttings								■	■	■		
	softwood cuttings			■	■	■							
Petroselinum crispum (parsley)	seeds sown directly				■	■	■	■					
Phacelia	seeds sown directly			■	■	■	■	■	■	■	■		
Philadelphus	hardwood cuttings	■	■									■	■
	semi-ripe cuttings								■	■	■		
	softwood cuttings			■	■	■							
Phlomis tuberosa	division			■	■	■							
	seeds sown under cover	■	■	■	■	■							
Phlox paniculata	division			■	■	■				■	■	■	
	semi-ripe cuttings								■	■	■		
	softwood cuttings			■	■	■							
Photinia x *fraseri*	hardwood cuttings	■	■									■	■
	semi-ripe cuttings								■	■	■		
Pittosporum tobira	air layering			■	■	■							
	layering			■	■	■							
	semi-ripe cuttings								■	■	■		
Platanus	hardwood cuttings	■	■									■	■
Polemonium	division			■	■	■							
	seeds sown under cover		■	■	■					■	■	■	
Polygonatum x *hybridum*	division			■	■	■							
Potentilla (perennial types)	division			■	■	■				■	■	■	
	semi-ripe cuttings								■	■	■	■	
Primula	division			■	■	■							
	seeds sown under cover			■	■	■	■			■	■	■	
Pulmonaria	division			■	■	■							

PLANT	Propagation method	Months											
		J	F	M	A	M	J	J	A	S	O	N	D
Puschkinia scilloides	division				■	■							
	offsets				■	■	■						
Pyracantha	layering			■	■	■							
	semi-ripe cuttings							■	■	■			
Rhododendron	air-layering			■	■	■							
	layering			■	■	■							
	semi-ripe cuttings								■	■	■		
Rhus typhina	semi-ripe cuttings								■	■	■		
Ribes sanguineum	hardwood cuttings	■	■									■	■
Rosa (rose)	hardwood cuttings	■	■									■	■
	semi-ripe cuttings								■	■	■		
	softwood cuttings			■	■	■							
Rudbeckia	division			■	■	■				■	■	■	
	seeds sown directly				■	■	■						
	seeds sown under cover		■	■	■								
Salix gracilistyla	hardwood cuttings	■	■									■	■
	softwood cuttings			■	■	■							
Salvia	division			■	■	■				■	■	■	
	semi-ripe cuttings								■	■	■		
	softwood cuttings			■	■	■							
Salvia rosmarinus (rosemary)	seeds sown under cover			■	■	■			■	■	■		
	semi-ripe cuttings								■	■	■		
	softwood cuttings			■	■	■							
Salvia yangii (syn. *Perovskia*)	seeds sown under cover	■	■	■	■	■	■	■	■	■	■	■	■
	semi-ripe cuttings								■	■	■		
	softwood cuttings			■	■	■							
Sambucus nigra	hardwood cuttings	■	■									■	■
	semi-ripe cuttings								■	■	■		
	softwood cuttings			■	■	■							
Sanguisorba	division			■	■	■			■	■	■		
Sarcococca	semi-ripe cuttings								■	■	■	■	
	softwood cuttings			■	■	■							

PLANT	Propagation method	Months											
		J	F	M	A	M	J	J	A	S	O	N	D
Satureja douglasii	layering			X	X	X	X	X	X				
	softwood cuttings			X	X	X	X	X	X	X			
Saxifraga	division			X	X	X							
Scabiosa	division		X	X	X								
	seeds sown under cover	X	X	X	X				X	X			
Scilla siberica	division				X	X							
	offsets					X	X	X					
Sedum	division			X	X	X			X	X	X		
Sempervivum	offsets			X	X	X			X	X			
Setaria italica	seeds sown directly					X	X						
	seeds sown under cover		X	X	X								
Skimmia	semi-ripe cuttings								X	X	X		
Spiraea	hardwood cuttings	X	X									X	X
	semi-ripe cuttings								X	X	X		
Stipa tenuissima	division			X	X	X							
	seeds sown under cover	X	X	X								X	X
Syringa	air layering			X	X	X			X	X	X		
	layering			X	X	X			X	X	X		
Taxus baccata	semi-ripe cuttings								X	X	X		
Tellima grandiflora	division			X	X	X							
	seeds sown under cover			X	X	X			X	X	X	X	
Thalictrum	division			X	X	X							
	seeds sown under cover		X	X	X				X	X			
Thunbergia alata	seeds sown under cover	X	X									X	X
Thymus (thyme)	division			X	X	X							
	seeds sown directly				X	X							
	seeds sown under cover		X	X	X								
Tiarella	division			X	X	X							
	seeds sown under cover			X	X	X				X	X	X	
Tropaeolum majus	seeds sown directly				X	X	X						
	seeds sown under cover			X	X								

PLANT	Propagation method	Months											
		J	F	M	A	M	J	J	A	S	O	N	D
Tulipa	division					■	■						
	offsets						■	■	■				
Verbascum	seeds sown directly									■	■		
	seeds sown under cover		■	■									
Verbena bonariensis	seeds sown directly					■	■						
	seeds sown under cover		■	■	■								
	semi-ripe cuttings								■	■	■		
	softwood cuttings			■	■	■							
Verbena officinalis	division			■	■	■							
	seeds sown under cover	■	■	■	■					■	■	■	
Veronica	basal cuttings			■	■	■							
	division			■	■	■				■	■		
	seeds sown under cover			■	■	■				■	■	■	
Veronicastrum	basal cuttings			■	■	■							
	division			■	■	■							
	seeds sown under cover			■	■	■				■	■	■	
Viburnum	layering			■	■	■							
	semi-ripe cuttings								■	■	■		
	softwood cuttings			■	■	■							
Viola	seeds sown directly				■	■	■			■	■		
	seeds sown under cover			■	■	■			■	■			
	softwood cuttings			■	■								
Visnaga daucoides	seeds sown directly			■	■					■	■		
Vitis vinifera	hardwood cuttings	■	■									■	■
Wisteria	layering			■	■	■				■	■	■	
Xerochrysum bracteatum	seeds sown under cover	■	■	■	■								
	softwood cuttings			■	■								

RESOURCES

Recommended books

The following books are a great investment for any money-conscious gardener. They are all well worth buying, but you could also see if your local library has a copy to borrow.

***RHS A–Z Encyclopedia of Garden Plants* edited by Christopher Brickell, 4th edition (DK, 2016)**
This truly is my bible. Every gardener should have this book. The most reliable source of information about garden plants.

***RHS Pruning and Training* by Christopher Brickell and David Joyce (DK, 2017)**
Everything you need to know about pruning over 800 plants.

***Containers in the Garden* by Claus Dalby (Cool Springs Press, 2022)**
This is a fantastic source of inspiration that can be easily translated into money-saving gardening. The use of perennials, even mixing them with ornamental vegetables, is truly innovative in this book.

***Compost* by Charles Dowding (DK, 2024)**
Learn from the best in the business. Compost is free to make, and, with guidance, simple to perfect.

***RHS Propagating Plants* edited by Alan Toogood (DK, 2019)**
This book is a must-have for anyone who would like to understand more about propagation, with over 1,500 different plants featured.

Websites

Royal Horticultural Society (RHS)
rhs.org.uk
There is no better place for me than the RHS website. Always up to date, it includes everything that a gardener needs.

Coppice Products
coppice-products.co.uk
This is a directory of coppice product makers in the UK, useful for finding wood for making plant supports.

Pinterest
pinterest.co.uk
I often use Pinterest in the winter just to feed my mind with ideas. It's a great place for money-saving gardeners who are looking for inspiration.

The Wildlife Trusts
wildlifetrusts.org
A great source of information on how to make your garden a wildlife-friendly place.

Butterfly Conservation
butterfly-conservation.org
A resource for learning, identifying, and understanding the most wonderful butterflies and moths, with ideas for how to provide what they need in your garden.

INDEX

Note: page numbers in **bold** refer to illustration captions and illustrations.

Author acknowledgments

I dedicate this book to my boys, Richard, William and Edward. You are my world.

I would also like to thank my mum who showed me how to find joy in simple things and who, by introducing me to propagation over 30 years ago, gave me a skill for life.

My very grateful thanks goes to Ruth O'Rourke for her guidance and support. I would also like to thank my fantastic editors Lucy Philpott and Zia Allaway, my designers Barbara Zuniga and Nikki Ellis, and to the most amazing team at DK: Maxine Pedliham, Silvia Dembner, and Hayley Reed. Thanks to my incredible photographer Britt Willoughby-Dyer, too. You are all simply the best!

I would like to thank all my social media followers for their ongoing support and positivity. Without you I wouldn't be where I am today and I hope that you will find this book useful.

Lastly, a big thank you to Chris Young for believing in me years ago.

Love Anya xxx

Publisher acknowledgments

DK would like to thank Adam Brackenbury for repro work, Kathy Steer for proofreading, Lisa Footitt for compiling the index, and Emily Cannings for editorial assistance.

The publisher would like to thank the following for their kind permission to reproduce their photographs:
(Key: a-above; b-below/bottom; c-centre; f-far; l-left; r-right; t-top)

Britt Willoughby Dyer: 11tl, 11bl, 12bl, 16, 19tl, 19bl, 20tl, 28, 30tl, 30tr, 30bl, 30br, 90tl, 124tl, 124bl, 124br, 142br, 145tr, 145bl, 150, 153tr, 153bl, 154tl, 154tr, 160, 163tl, 163tr, 164tl, 164tr; **Anya Lautenbach:** 134, 164bl, 167br

Editorial Manager Ruth O'Rourke
Project Editor Lucy Philpott
Senior Designer Barbara Zuniga
Production Editor David Almond
Senior Production Controller Stephanie McConnell
DTP and Design Coordinator Heather Blagden
Sales Material & Jackets Coordinator Emily Cannings
Art Director Maxine Pedliham

Editorial Zia Allaway
Design Nikki Ellis
Photography Britt Willoughby
Consultant Gardening Publisher Chris Young

First published in Great Britain in 2024 by
Dorling Kindersley Limited
DK, 20 Vauxhall Bridge Road,
London SW1V 2SA

The authorised representative in the EEA is
Dorling Kindersley Verlag GmbH. Arnulfstr. 124,
80636 Munich, Germany

A Penguin Random House Company
10 9 8 7 6
006–348763–Oct/2024

A CIP catalogue record for this book is available from the British Library.

ISBN: 978-0-2417-3306-6

Printed and bound in Latvia
www.dk.com

Anya Lautenbach is a self-taught gardener with a passion for the environment and championing neurodiversity. Her first book, *The Money-Saving Gardener*, was a *Sunday Times* Bestseller; in her second book she guides you through each month of the year, with growing advice, money-saving tips, and moments to be mindful.

She grew up in Poland, and after travelling for many years in Germany and the Scottish Highlands, she now lives with her husband and two sons in Buckinghamshire, UK, where her garden has blossomed through years of propagation and clever gardening tricks.

Anya's lifelong passion for nature and for propagating plants inspired her to share her knowledge to her followers across social media. There, as Anya the Garden Fairy, she provides easy-to-follow tutorials covering a range of accessible and achievable gardening techniques, and reveals tips and tricks for creating high-impact, low-cost gardens that work in harmony with nature. Anya also uses her platform to raise awareness about neurodiversity and the positive influence that horticulture can have on mental health. She proves that gardening does not need expensive equipment or specialist training – anyone can transform their garden and create something truly amazing.

You can find Anya online at:

Instagram: @anya_thegarden_fairy
Facebook: @anya_thegarden_fairy
Tiktok: @anyathegardenfairy
Youtube: @anyathegardenfairy
Website: www.anyalautenbach.com